T0342550

Applied Smart Health Care Informatics

Series Preface Dr. Siddhartha Bhattacharyya, CHRIST (Deemed to be University), Bengaluru, India (Series Editor)

The Intelligent Signal and Data Processing (ISDP) book series is aimed at fostering the field of signal and data processing, which encompasses the theory and practice of algorithms and hardware that convert signals produced by artificial or natural means into a form useful for a specific purpose. The signals might be speech, audio, images, video, sensor data, telemetry, electrocardiograms, or seismic data, among others. The possible application areas include transmission, display, storage, interpretation, classification, segmentation, or diagnosis. The primary objective of the ISDP book series is to evolve future-generation scalable intelligent systems for faithful analysis of signals and data. ISDP is mainly intended to enrich the scholarly discourse on intelligent signal and image processing in different incarnations. ISDP will benefit a wide range of learners, including students, researchers, and practitioners. The student community can use the volumes in the series as reference texts to advance their knowledge base. In addition, the monographs will also come in handy to the aspiring researcher because of the valuable contributions both have made in this field. Moreover, both faculty members and data practitioners are likely to grasp depth of the relevant knowledge base from these volumes.
The series coverage will contain, not exclusively, the following:

1. **Intelligent signal processing**
   a) Adaptive filtering
   b) Learning algorithms for neural networks
   c) Hybrid soft-computing techniques
   d) Spectrum estimation and modeling
2. **Image processing**
   a) Image thresholding
   b) Image restoration
   c) Image compression
   d) Image segmentation
   e) Image quality evaluation
   f) Computer vision and medical imaging
   g) Image mining
   h) Pattern recognition
   i) Remote sensing imagery
   j) Underwater image analysis
   k) Gesture analysis
   l) Human mind analysis
   m) Multidimensional image analysis
3. **Speech processing**
   a) Modeling
   b) Compression
   c) Speech recognition and analysis
4. **Video processing**
   a) Video compression
   b) Analysis and processing
   c) 3D video compression
   d) Target tracking
   e) Video surveillance
   f) Automated and distributed crowd analytics
   g) Stereo-to-auto stereoscopic 3D video conversion
   h) Virtual and augmented reality
5. **Data analysis**
   a) Intelligent data acquisition
   b) Data mining
   c) Exploratory data analysis
   d) Modeling and algorithms
   e) Big data analytics
   f) Business intelligence
   g) Smart cities and smart buildings
   h) Multiway data analysis
   i) Predictive analytics
   j) Intelligent systems

Applied Smart Health Care Informatics

A Computational Intelligence Perspective

Edited by

Sourav De
Cooch Behar Govt. Engineering College, Cooch Behar, West Bengal, India

Rik Das
Xavier Institute of Social Service, Ranchi, Bihar, India

Siddhartha Bhattacharyya
CHRIST (Deemed to be University), Bengaluru, India

Ujjwal Maulik
Jadavpur University, Kolkata, West Bengal, India

This edition first published 2022

Registered Offices
John Wiley & Sons, Inc., 111 River Street, Hoboken, NJ 07030, USA
John Wiley & Sons Ltd, The Atrium, Southern Gate, Chichester, West Sussex, PO19 8SQ, UK

Editorial Office
The Atrium, Southern Gate, Chichester, West Sussex, PO19 8SQ, UK

For details of our global editorial offices, customer services, and more information about Wiley products visit us at www.wiley.com.

Wiley also publishes its books in a variety of electronic formats and by print-on-demand. Some content that appears in standard print versions of this book may not be available in other formats.

Library of Congress Cataloging-in-Publication Data

Names: De, Sourav, 1979- editor. | Das, Rik, 1978- editor. | Bhattacharyya, Siddhartha, 1975- editor. | Maulik, Ujjwal, editor.
Title: Applied smart health care informatics : a computational intelligence perspective / Sourav De, Rik Das, Siddhartha Bhattacharyya, Ujjwal Maulik.
Description: First edition. | Hoboken, NJ : Wiley, 2022. | Series: The Wiley series in intelligent signal and data processing | Includes index.
Identifiers: LCCN 2021039430 (print) | LCCN 2021039431 (ebook) | ISBN 9781119743170 (hardback) | ISBN 9781119743200 (adobe pdf) | ISBN 9781119742982 (epub)
Subjects: LCSH: Medical informatics.
Classification: LCC R858 .A69 2022 (print) | LCC R858 (ebook) | DDC 610.285–dc23/eng/20211001
LC record available at https://lccn.loc.gov/2021039430
LC ebook record available at https://lccn.loc.gov/2021039431

Cover Design: Wiley
Cover Image: © sdecoret/Shutterstock

Set in 9.5/12.5pt STIXTwoText by Straive, Chennai, India
Printed and bound by CPI Group (UK) Ltd, Croydon, CR0 4YY

C9781119743170_160222

Sourav De would like to dedicate this book to his respected father-in-law,
Late Barun Ghosh.

Rik Das would like to dedicate this book to his father Mr. Kamal Kumar Das,
his mother Mrs. Malabika Das, his wife Mrs. Simi Das, and
his children Sohan and Dikshan.

Siddhartha Bhattacharyya would like to dedicate this book to Late Nalini Bhuson
Acharjee and Late Nirmala Acharjee, the parents-in-law of his second eldest sister.

Ujjwal Maulik would like to dedicate this volume to all his students,
including his son UTSAV, his colleagues, and his collaborators.

Contents

Preface

Health care informatics, aka medical informatics, refers to the application of information engineering and management to the field of health care, which covers the management and use of patient health care information. By means of a multi-disciplinary approach, it uses health information technology to improve health care by migrating to newer and higher quality opportunities. The United States National Library of Medicine (NLM) defines health informatics as "an interdisciplinary study of the design, development, adoption and application of IT-based innovations in health care services delivery, management and planning." Essentially, it affects the optimization of the acquisition, storage, retrieval, and use of information in health and bio-medicine. Intelligent health care informatics augments the purview of existing health care amenities by adapting intelligent technologies to information engineering. Intelligent analysis of the information therein enhances the overall management as far as resource use is concerned.

With the advent of Big Data analysis, intelligent health care informatics has called for the efficient and effective use of healthcare data and the diagnosis thereof. During the next few years, there must be a sea change in the approaches to health care management. Smart pills may come to the foray as Bio-MEMS drug delivery systems or intelligent drug delivery systems. Wearable medical devices could be attached to the patient's body to keep in touch with physicians for real time monitoring. Nano-bots might be used to collect specimens or look for early signs of disease. Content management could also become more intelligent and intricate.

Patients with chronic disease live for decades through modern medication, surgery, close supervision, and other modern treatments. Soon, patients can manage their healthcare conditions. They can also take necessary measures to prevent escalation and deterioration of their health. Curative and reactive healthcare approaches will switch to preventive and proactive health management. Someday, people will be able to control their own lifestyle and future health, and that will bring a revolution.

In this journey, artificial intelligence or computational intelligence will play a pivotal role in improving the quality of services of healthcare systems, and that will bring a better coordination of care. Intelligent health will be the potential solution to keep up with the escalating increase of healthcare cost. Huge amounts of existing data in the healthcare sector can be managed with the tools of intelligent systems like machine learning, meta-heuristic algorithms, big data, deep learning, internet-of-things (IoT), etc. It will be easy and faster for the surgeons, hospital, medical, and emergency staff to find the probable treatment or drug for rare diseases. Innovations of the intelligent systems in the healthcare arena may help society by reducing the cost and time of medical treatments; concrete solutions for a particular disease can be easily found.

This volume, comprising eight well-versed chapters (apart from the introductory and concluding chapters), will entice the readers to engage with major emerging trends in technology that are supporting the advancement of the medical image analysis with the help of artificial intelligence and computational intelligence. This volume elaborates on the fundamentals and advancement of conventional approaches in the field of health care management. The scope of this volume also opens an arena in which researchers propose new approaches and review state-of-the-art machine learning, computer vision, and soft computing techniques as well as relate the same to their applications in medical image analysis. The motivation of this volume is not only to put forward new ideas in technology innovation but also to analyse the effect of the same in the current context of medical healthcare.

Health care informatics, also referred as biomedical or medical informatics, is an application of information engineering and management in the medical field. Health care fundamentally covers the management and employment of patient health care information. It is a multidisciplinary field that studies and pursues the effectual use of biomedical data, knowledge for scientific inquiry, information, problem solving, and decision making. Chapter 1 provides an overview of a few smart healthcare practices.

Lung cancer is a fatal form of cancer around the world. The American Lung Association reports an estimated five-year survival rate in lung cancer patients of 18.6%. The statistics affirm that the survival rate is significantly lower than in other forms of cancer. However, the five-year survival rate stands at 56% when the disease is diagnosed in a localized stage. Some cases do not appear to have symptoms until cancer has reached a later stage. The primary cause of concern is the low percent of early lung cancer detection, which is merely 16%. Lung cancer staging is a procedure associated with the disease's successful prognosis and formulation of an efficient treatment plan. Medical imaging techniques play a vital role in the diagnosis of lung cancer. Accuracy is crucial in treatment as lung cancer is influenced by internal and external factors or mistaken for other pulmonary diseases. The staging of cancer allows for the significant elimination of treatment failures. However, cancer staging is a dynamic process that involves multiple and frequent modifications to recognize organ features. The staging process requires a more robust and automated technique that can provide sensitive and unique input to improve the overall treatment process. Thus, artificial intelligence sub-branches such as deep learning play a vital role in initiating such improvements for an efficient cancer staging process. Chapter 2 uncovers the potential of a deep learning model combined with positron emission tomography—computed tomography (PET-CT) to develop a technique that identifies tumors with more precision. The proposed research will assist doctors in accurately measuring the tumor and identifying the stage of lung cancer that will determine further treatment and an exact prognosis.

Cyber-physical attacks (CP attacks), originating in cyber space but damaging physical infrastructure, are a significant recent research focus. Such attacks have affected many cyber-physical systems (CPSs) such as smart grids, intelligent transportation systems, and medical devices. In Chapter 3, the authors consider techniques for the detection and mitigation of CP attacks on medical devices. It is obvious that such attacks have immense safety implications. This work is based on formal methods, a class of mathematically founded techniques for the specification and verification of safety-critical systems. The interaction of a cardiac pacemaker is discussed. Subsequently, the authors provide an overview of formal methods with particular emphasis on run-time based approaches, which are ideal for the design of security monitors. Two recently developed approaches are illustrated that assist in the detection of attacks as well as mitigation.

Integrating heterogeneous omics data profiles, such as genomics, epigenomics, and transcriptomics may provide new insights into discovering some unknown genomic mechanisms involved in cancer and other related complex diseases. The alterations of multiple omics, including gene mutations, epigenetic changes, and gene regulation modifications, are responsible for tumor initiation and cancer progression. Most of the multi-view data profiles contain a huge number of genes, many of which are redundant, noisy, and irrelevant. It is computationally impractical to use these massive data sets without any filtering of the feature set. High performance (deep) machine learning strategies now appear to be an essential tool to learn the hidden structure from the data. In Chapter 4, the authors have proposed a two-step approach to systematically identify gene signatures from multi-omics head and neck cancer data. First, an autoencoder-based strategy is used to integrate gene expression and methylation data. From this, the features are extracted by using the information from the bottleneck layer of the autoencoder. The features represent the combined representation of the two omics profiles. Next, the features that stem from the integrated data are applied to learn another deep learning model called the capsule network. The coupling coefficients between primary and output capsules are also analysed to interpret the features captured by the capsules.

The last two decades have witnessed unprecedented advancements in computational techniques and artificial intelligence. These new developments are going to greatly impact biological data analysis for the health care system. In fact, the availability of large scale high-throughput biomedical data sets offers a fertile ground for application of these AI-based techniques in to extract valuable information that can be harnessed in the diagnosis and treatment of various diseases. Chapter 5 provides a comprehensive review of computational tools and online resources for high throughput analyses of biomedical data. It focuses on single-cell RNA sequencing data, multi-omics data integration, drug design with AI, medical imaging data analysis, and IoT. After providing a brief overview of the fundamental biological terms, a variety of research problems are described in the health care system and how various high throughput data can help solving them. Next, an in depth overview of machine learning techniques of computing and learning methods that can be used in a variety of sequencing data analyses is provided.

Cancer is one of the most devastating diseases worldwide. It affects nearly every household, although cancer types are prevalent in different geographical regions. One example is breast cancer, which is the most common type of cancer in women worldwide. Therefore, prevention strategies are needed to address this issue. Identifying risk factors of breast cancer is crucial since it allows physicians to acquaint them with the risks. Accordingly, physicians can recommend precautionary actions. In the first part of Chapter 6, the authors detail the discovery of significant rules for breast cancer patients, focusing on different ethnic groups. Predicting the risk of the occurrence of breast cancer is an essential issue for clinical oncologists. A reliable prediction will help oncologists and other clinicians in their decision-making process and allow clinicians to choose the most reliable and evidence-based treatment. In the second part of the chapter, a super learner or stacked ensemble technique is employed to the breast cancer data set obtained from the Breast Cancer Surveillance Consortium (BCSC) database. A comparison of the performance of the super learner and the individual base learners is conducted. The results of the first part of this study (rule extraction from breast cancer patients in distinct ethnic groups) found well-known ethnic disparities in cancer prevalence. The experimental results revealed that the produced rules hold the highest confidence level. The crucial rules, which can be easily understood, are also interpreted.

Negative-stain transmission electron microscopy (TEM) is considered a fundamental approach for virus detection and identification. In this context, Chapter 7 presents a new architecture, based on neuro-rough hybridization, for the analysis of TEM images. It assumes that a specific local descriptor at a given scale may be relevant in classifying a particular pair of virus classes but may not be able to encapsulate the inherent characteristics of another pair of classes. Important features from class-pair relevant descriptors are, therefore, first identified using the rough hypercuboid approach, and then discriminatory features are learned using the contrastive divergence algorithm of the restricted Boltzmann machine (RBM). Finally, a support vector machine (SVM) with a linear kernel is adopted to categorize the TEM images into one of the known virus classes. The proficiency of the proposed approach with respect to several state-of-the-art methods was established on a publicly available, benchmark Virus data set.

Computer vision plays a substantial role in health care applications such as the diagnosis of diseases and planning for treatment. Brain tumors are severe conditions that may be deadly if not detected and treated early. In India, brain tumors occur in five to ten people per one lakh population (100 000 people). Deep learning is a category of artificial intelligence that does not require any human intervention to learn the features. Deep learning algorithms learn the features of images on their own and are capable of learning more complex features from the images. While characterizing the deep neural networks, the selection of optimizers plays a vital role. Optimizers are used to minimize the loss function by varying the weights and learning rate attributes of the neural network. Optimization algorithms are essential for producing more accurate results by reducing the loss function of the neural network. In Chapter 8, the authors have analyzed popular optimizers such as sgd, adam, rmsprop, adagrad, adadelta, adamax, and nadam used with artificial neural network systems in the proposed work. Two models, a simple artificial neural network (ANN) model and a convolutional neural network (CNN) model, have been considered. Each optimizer is executed with these two models to classify abnormal slices from magnetic resonance imaging (MRI) of human brain scans. The BraTS2013 and WBA data sets were used for training and testing the models. The accuracies of every model were recorded to analyse the optimizer's performance.

In Chapter 9, a machine learning approach is proposed to predict whether the given brain MRI scans are normal or abnormal. This prediction is needed for treatment planning and diagnosis. The proposed method makes use of the bilateral symmetric nature of the human brain by splitting it into the left and right hemispheres (LHS and RHS) to extract the feature differences between the hemispheres. A feature set of 763 x 39 dimensions are created as the input for the classification model. Among these 39 features, 16 were selected by the Pearson's correlation coefficient to have correlation value greater than 0.3. To train the model, six tumor volumes from the BraTS2013 and two normal volumes from the IBSR-18 data sets were used. For testing the model, 11 tumor volumes from the BraTS2013 and two normal volumes from the IBSR-18 data sets were used. The k-nearest neighbourhood (KNN) model was trained using the training data and the prediction done on the test data. A stratified k-fold cross-validation was used to validate the proposed model. The proposed model was analysed in terms of false alarm (FA), missed alarm (MA), and accuracy (ACC) for performance. The results showed that the proposed model yielded a 98 and 95.6% accuracy on the validation and testing data, respectively.

Chapter 10 draws a line of conclusion on the future aspects of healthcare informatics while stressing the need for the effective management of healthcare resources.

This volume will benefit several categories of students and researchers. At the student level, this volume can serve as a treatise/reference book for the special papers at the master's level aimed at inspiring future researchers. Newly inducted PhD aspirants would also find the contents of this volume useful as far as their compulsory coursework is concerned. At the researchers' level, those interested in interdisciplinary research would also benefit from the volume. After all, the enriched interdisciplinary contents of the volume will always be a subject of interest to the faculties, existing research communities, and new research aspirants from diverse disciplines of the concerned departments of premier institutes across the globe.

Cooch Behar, India	*Sourav De*
Ranchi, India	*Rik Das*
Bengaluru, India	*Siddhartha Bhattacharyya*
Kolkata, India	*Ujjwal Maulik*
December, 2021	

About the Editors

Dr. Sourav De completed his bachelor's in information technology at The University of Burdwan, Burdwan, India in 2002. He earned his master's in information technology from The West Bengal University of Technology, Kolkata, India in 2005. He completed his PhD in Computer Science and Technology at the Indian Institute of Engineering & Technology, Shibpur, Howrah, India in 2015. He is an Associate Professor in the Computer Science & Engineering Department at the Cooch Behar Government Engineering College, West Bengal. Before 2016, he was an Assistant Professor for more than ten years in the Department of Computer Science and Engineering and Information Technology of the University Institute of Technology, The University of Burdwan, Burdwan, India. He served as a Junior Programmer at the Apices Consultancy Private Limited, Kolkata, India in 2005. He is a co-author of one book, the co-editor of twelve books, and has more than 54 research publications in internationally reputed journals, international books, and international IEEE conference proceedings as well as five patents in his name. Dr. De has served as a reviewer for several international IEEE conferences and on several international editorial books. He has also served as a reviewer at reputed international journals such as Applied Soft Computing, Elsevier, BV, Knowledge-Based Systems, Computer Methods in Biomechanics and Biomedical Engineering Imaging & Visualization, Inderscience Journals, etc. He has been a member of organizing and technical program committees for several national and international conferences and been invited to seminars as an expert speaker. His research interests include soft computing, pattern recognition, image processing, and data mining. Dr. De is a Senior member of IEEE and a member of the ACM, Institute of Engineers (IEI), Computer Science Teachers Association (CSTA), Institute of Engineers, and IAENG, Hong Kong. He is a life member of ISTE, India.

Dr. Rik Das is an Assistant Professor in the Post Graduate Programme in Information Technology, at the Xavier Institute of Social Service, Ranchi. He has a PhD (Tech) in Information Technology from the University of Calcutta. He has also received his MTech (Information Technology) from the University of Calcutta after earning his BE (Information Technology) from the University of Burdwan. Dr. Das has over 16 years of experience in academia and research with several leading universities and institutes in India including the Narsee Monjee Institute of Management Studies (NMIMS) (Deemed-to-be-University), Globsyn Business School, Maulana Abul Kalam Azad University of Technology, and so on. He had an early career stint in business development and project marketing with industries like Great Eastern Impex Pvt. Ltd. and Zenith Computers Ltd. Dr. Rik Das was appointed as a "Distinguished Speaker" by the Association of Computing Machinery (ACM) in July 2020. He was featured on the uLektz Wall of Fame as one of the "Top 50 Tech Savvy Academicians in Higher Education across India" for the year 2019. He is also a member of the International Advisory Committee of AI-Forum, UK. Dr. Das was awarded a professional membership to the ACM for the year 2020-21. In 2020, he was also the recipient of prestigious "InSc Research Excellence Award". Dr. Das was conferred with the "Best Researcher Award" at the International Scientist Awards on Engineering, Science and Medicine for the year 2021. He is also the recipient of the 2021 "Best Innovation Award" in the Computer Science category of the UILA Awards. Dr. Das has conducted collaborative research with professionals from industries including Philips-Canada, Cognizant Technology Solutions, and TCS. His keen interest toward the application of machine learning and deep learning techniques for designing computer-aided diagnosis systems has resulted in joint research publications with professors and researchers from several universities abroad including the College of Medicine, University of Saskatchewan, Canada; Faculty of Electrical Engineering and Computer Science, VSB Technical University of Ostrava, Ostrava, Czechia; and Cairo University, Giza, Egypt.

Dr. Das has filed and published two Indian patents, consecutively during 2018 and 2019 and has over 40 international publications to-date with reputed publishers including IEEE, Springer, Emerald, and Inderscience. He has also authored three books in the domain of content-based image classification and has edited three volumes to-date with IGI Global, CRC Press, and De Gruyter, Germany. Dr. Das has chaired several sessions in international conferences on artificial intelligence and machine learning as a domain expert. He has also served as an invited speaker at several national and international technical events, conclaves, meetups, and refresher courses on data analytics, artificial intelligence, machine learning, deep learning, image processing, and e-learning, which have been organized and hosted by prominent bodies like the University Grants Commission (Human Resource Development Centre), the Confederation of Indian Industry (CII), software consulting organizations, the MHRD Initiative under the Pandit Madan Mohan Malviya National Mission on Teachers and Teaching, IEEE student chapters, and the computer science/information technology departments of leading universities. Dr. Rik Das's hobby is a YouTube channel called 'Curious Neuron', where he discusses the latest knowledge and information to larger communities in the domain of machine learning, research and development, as well as open-source programming languages. Dr. Das is always open to discuss new research and project ideas for collaborative work or techno-managerial consultancies.

Dr. Siddhartha Bhattacharyya earned his bachelors in physics, bachelors in optics and optoelectronics, and masters in optics and optoelectronics from the University of Calcutta, India in 1995, 1998, and 2000, respectively. He completed a PhD in

Computer Science and Engineering at Jadavpur University, India in 2008. He is the recipient of the University Gold Medal from the University of Calcutta for his masters. Dr. Bhattacharyya is also the recipient of several coveted awards including the "Distinguished HoD" and "Distinguished Professor" awards conferred by the Computer Society of India, Mumbai Chapter, India in 2017; the "Honorary Doctorate Award" (DLitt) from The University of South America; and the South East Asian Regional Computing Confederation (SEARCC) "International Digital Award ICT Educator of the Year" in 2017. He was appointed an ACM "Distinguished Speaker" for the 2018 to 2020 term and been inducted into the "People of ACM Hall of Fame" in 2020. Dr. Bhattacharyya has been appointed an IEEE Computer Society "Distinguished Visitor" for the 2021 to 2023 term and elected a full foreign member of the Russian Academy of Natural Sciences. Dr. Bhattacharyya is the Principal of Rajnagar Mahavidyalaya, Rajnagar, Birbhum. He served as a Professor in the Department of Computer Science and Engineering at Christ University, Bangalore after he served as the Principal of the RCC Institute of Information Technology, Kolkata, India from 2017 to 2019. He has also served as a Senior Research Scientist in the Faculty of Electrical Engineering and Computer Science at the VSB Technical University of Ostrava, Czech Republic (2018 to 2019). Prior to this, he was the Professor of Information Technology at the RCC Institute of Information Technology, Kolkata, India. He served as the head of the department from March 2014 to December 2016. Prior to this, he was an Associate Professor of Information Technology at the RCC Institute of Information Technology, Kolkata, India from 2011 to 2014. From 2005 to 2011, Dr. Bhattacharyya was an Assistant Professor in Computer Science and Information Technology at the University Institute of Technology, The University of Burdwan, India. He was a Lecturer in Information Technology at the Kalyani Government Engineering College, India from 2001 to 2005. Dr. Bhattacharyya is a co-author of six books, the co-editor of 75 books, and has more than 300 research publications in international journals and conference proceedings as well as two PCTs (Patent Cooperation Treaty). He has been a member of organizing and technical program committees for several national and international conferences. He was the Founding Chair of ICCICN 2014, ICRCICN (2015 to 2018), and ISSIP (2017, 2018) (Kolkata, India). He was also the General Chair of several international conferences including WCNSSP 2016 (Chiang Mai, Thailand), ICACCP (2017, 2019; Sikkim, India), and ICICC (2018; New Delhi, India and 2019; Ostrava, Czech Republic).

Dr. Bhattacharyya is the Associate Editor of several reputed journals including Applied Soft Computing, IEEE Access, Evolutionary Intelligence, and IET Quantum Communications. He is the editor of the International Journal of Pattern Recognition Research and the Founding Editor in Chief of the International Journal of Hybrid Intelligence, Inderscience. He has guest-edited several issues with several international journals and is serving as the Series Editor of the IGI Global Book Series "Advances in Information Quality and Management (AIQM)", De Gruyter Book Series "Frontiers in Computational Intelligence (FCI)", CRC Press Book Series(s) "Computational Intelligence" and "Applications & Quantum Machine Intelligence", Wiley Book Series "Intelligent Signal and Data Processing",

Elsevier Book Series "Hybrid Computational Intelligence for Pattern Analysis and Understanding", and Springer "Tracts on Human Centered Computing".

His research interests include hybrid intelligence, pattern recognition, multimedia data processing, social networks, and quantum computing.

Dr. Bhattacharyya is a life fellow of the Optical Society of India (OSI), India; life fellow of the International Society of Research and Development (ISRD), UK; a fellow of the Institution of Engineering and Technology (IET), UK; a fellow of the Institute of Electronics and Telecommunication Engineers (IETE), India; and a fellow of the Institution of Engineers (IEI), India. He is also a senior member of the Institute of Electrical and Electronics Engineers (IEEE), USA; International Institute of Engineering and Technology (IETI), Hong Kong; and the Association for Computing Machinery (ACM), USA. He is a life member of the Cryptology Research Society of India (CRSI), Computer Society of India (CSI), Indian Society for Technical Education (ISTE), Indian Unit for Pattern Recognition and Artificial Intelligence (IUPRAI), Center for Education Growth and Research (CEGR), Integrated Chambers of Commerce and Industry (ICCI), and Association of Leaders and Industries (ALI). He is a member of the Institution of Engineering and Technology (IET), UK; International Rough Set Society; International Association for Engineers (IAENG), Hong Kong; Computer Science Teachers Association (CSTA), USA; International Association of Academicians, Scholars, Scientists and Engineers (IAASSE), USA; Institute of Doctors Engineers and Scientists (IDES), India; The International Society of Service Innovation Professionals (ISSIP); and The Society of Digital Information and Wireless Communications (SDIWC). He is also a certified Chartered Engineer of the IEI and is on the Board of Directors of IETI, Hong Kong.

Dr. Ujjwal Maulik has been a Professor in the Department of Computer Science and Engineering, Jadavpur University since 2004. He formerly acted as the Head of the same department. He held the position of the Principal and Head of the Department of Computer Science and Engineering. Dr. Maulik has worked in many universities and research laboratories around the world as a visiting Professor/Scientist including the Los Alamos National Lab, USA in 1997; University of New South Wales, Australia in 1999; University of Texas at Arlington, USA in 2001; University of Maryland at Baltimore County, USA in 2004; Fraunhofer Institute for Autonome Intelligent Systems, St. Augustin, Germany in 2005; Tsinghua University, China in 2007; Sapienza University, Rome, Italy in 2008; University of Heidelberg, Germany in 2009; German Cancer Research Center (DKFZ), Germany in 2010, 2011, and 2012; Grenoble INP, France in 2010, 2013, and 2016; University of Warsaw in 2013 and 2019; University of Padova, Italy in 2014 and 2016; Corvinus University, Budapest, Hungary in 2015 and 2016; University of Ljubljana, Slovenia in 2015 and 2017; and International Center for Theoretical Physics (ICTP), Trieste, Italy in 2014, 2017, and 2018. He is the recipient of the BOYSCAST Fellowship from the Government of India in 2001, the Alexander von Humboldt Fellowship from 2010 to 2012, and the Senior Associate of ICTP, Italy from 2012 to 2018. He is a Fellow of the Institute of Electrical and Electronics Engineers (IEEE), USA;

Indian National Academy of Engineers (INAE), India; International Association for Pattern Recognition (IAPR), USA; West Bengal Academy of Science and Technology (WAST), India; Institution of Engineers (IE), India; and Institution for Electronics and Telecommunication Engineers (IETE), India, 2001. He is also an ACM "Distinguished Speaker". His research interests include machine learning, pattern analysis, data science, bioinformatics, multi-objective optimization, social networking, IoT, and autonomous car. In these areas, he has published ten books, more than 350 papers, filed several patents, and mentored 20 doctoral students. His other interests include outdoor sports, music, and traveling extensively around the world.

List of Contributors

Allen, Nathan
University of Auckland
New Zealand

Bandyopadhyay, Sanghamitra
Machine Intelligence Unit
Indian Statistical Institute
Kolkata
India

Bhadra, Tapas
Department of Computer Science and Engineering
Aliah University
Kolkata
Newtown
India

Bhattacharyya, Siddhartha
CHRIST (Deemed to be University)
Bengaluru
India

Choudhuri, Soham
Computational Natural Sciences and Bioinformatics
International Institute of Information Technology
Hyderabad
Gachibowli
India

Das, Kaushik
Department of Computer Science
CHRIST (Deemed to be University)
Bangalore
India

Das, Rik
Department of Information Technology
Xavier Institute of Social Service
Ranchi
Jharkhand
India

De, Sourav
Department of Computer Science & Engineering
Cooch Behar Government Engineering College
Cooch Behar
West Bengal
India

Gaur, Manoj Singh
Indian Institute of Technology Jammu
India

Ghosh, Bhaswar
Computational Natural Sciences and Bioinformatics
International Institute of Information Technology
Hyderabad
Gachibowli
India

J, Chandra
Department of Computer Science
CHRIST (Deemed to be University)
Bangalore
India

Kabir, Faisal
Department of Computer Science
Pennsylvania State
University-Harrisburg
PA, USA

Kalaichelvi, Nagarajan
Department of Computer Science and
Applications
The Gandhigram Rural Institute
(Deemed to be University)
Dindigul
Tamilnadu
India

Kalaiselvi, Thiruvenkadam
Department of Computer Science and
Applications
The Gandhigram Rural Institute
(Deemed to be University)
Dindigul
Tamilnadu
India

Kumar, Debamita
Biomedical Imaging and Bioinformatics
Lab
Machine Intelligence Unit
Indian Statistical Institute
Kolkata
India

Li, Aimin
Shaanxi Key Laboratory for Network
Computing and Security Technology
School of Computer Science and
Engineering
Xi'an University of Technology
Shaanxi
Xi'an
China

Ludwig, Simone
Department of Computer Science
North Dakota State University
ND, USA

M, Nachamai
Siemens Healthcare Pvt. Ltd
Bangalore
Karnataka
India

Maji, Pradipta
Biomedical Imaging and Bioinformatics
Lab
Machine Intelligence Unit
Indian Statistical Institute
Kolkata
India

Mallik, Saurav
Center for Precision Health
School of Biomedical Informatics
University of Texas Health Science
Center at Houston
TX, USA

Maulik, Ujjawal
Department of Computer Science and Engineering
Jadavpur University
West Bengal
Kolkata
India

Padmapriya, Thiyagarajan
Department of Computer Science and Applications
The Gandhigram Rural Institute (Deemed to be University)
Dindigul
Tamilnadu
India

Pearce, Hammond
NYU Tandon School of Engineering
NY
USA

Pinisetty, Srinivas
Indian Institute of Technology
Bhubaneswar
India

Ray, Sumanta
Department of Computer Science and Engineering
Aliah University
Kolkata
India

Roop, Partha
University of Auckland
New Zealand

Saha, Suparna
SyMeC Data Center
Indian Statistical Institute
Kolkata
India

Si, Tapas
Department of Computer Science and Engineering
Bankura Unnayani Institute of Engineering
Subhankar Nagar
Pohabagan
India

Somasundaram, Karuppanagounder
Department of Computer Science & Application
The Gandhigram Rural Institute (Deemed to be University)
Dindigul
Tamilnadu
India

Trew, Mark
University of Auckland
New Zealand

1

An Overview of Applied Smart Health Care Informatics in the Context of Computational Intelligence

Sourav De[1,] and Rik Das[2]*

[1]*Department of Computer Science & Engineering, Cooch Behar Government Engineering College, Vill- Harinchawra, P.O.- Ghughumari, Cooch Behar, West Bengal, 736170, India*
[2]*Department of Information Technology, Xavier Institute of Social Service, Post Box No-7, Dr Camil Bulcke Path, Ranchi, Jharkhand, 834001, India*

1.1 Introduction

Health care informatics, in other words medical informatics, alludes to the use of data design and onboarding to the field of medical care, which basically covers the administration and utilization of patient medical services data. Through a multidisciplinary approach, it utilizes health information technology to improve medical care by depending on more advanced opportunities. According to the United States National Library of Medicine (NLM), health informatics is "an interdisciplinary study of the design, development, adoption and application of IT-based innovations in health care services delivery, management and planning" (DeBakey, 1991). Basically, it impacts the improvement of the obtaining, stockpiling, recovery, and utilization of data in health and bio-medication. Intelligent health care informatics expand the domain of current medical care conveniences by encompassing aspects from intelligent technologies to computational engineering. Intelligent analysis of the information upgrades the general administration by taking everything into account.

Health care informatics combine the fields of information technology, science, and medicine for a better seamless and speedy management process that serves people worldwide. The main objective for health care informatics is to render effective health care to patients with the help of technologic advancements in public health, drug discovery, pharmacy, etc. However, there is an insufficient understanding of the computational methodologies that will be highly efficient for the health care sector and its approach for patients worldwide (Durcevic, 2020).

Belle et al. (2015) discussed various smart health care informatics that can be tackled using computational techniques. Big data analytics is required for the health care sector due to the rising costs in nations like the United States (Durcevic, 2020).

*Corresponding Author: Sourav De; dr.sourav.de79@gmail.com

Applied Smart Health Care Informatics: A Computational Intelligence Perspective, First Edition.
Edited by Sourav De, Rik Das, Siddhartha Bhattacharyya, and Ujjwal Maulik.

Moreover, expenses are much higher than they ought to be, and they have been rising over the last 20 years. Distinctly, we are in need of smart, data-driven improvements in the health care sector.

1.2 Big Data Analytics in Healthcare

Big data analytics can be applied in different areas of medicine. The concept of big data analytics can be employed in different areas and among them image processing, signal processing, and genomics (Ritter et al., 2011) are primarily noted.

Medical images are a vital source of data, and they are frequently employed for diagnosing, assessing therapy, and designing (Ritter et al., 2011) algorithms. X-ray, magnetic resonance imaging (MRI), molecular imaging, computerized tomography (CT) images, photo acoustic imaging, ultrasound, fluoroscopy, and mammography are some instances of imaging methods that are found inside clinical settings (Belle et al., 2015). Medical images can run from a couple of megabytes to process a solitary report to many megabytes per analysis: for example, thin-slice CT studies (Seibert, 2010). These types of information need huge capacity limits for long term data retention and require accurate and fast algorithms for decision-assisting automation. Likewise, if other sources of information obtained for an individual patient are additionally applied at diagnoses, prognosis, and treatment, then the issue of proving reliable storage and increasingly advantageous methods for this large scope of records turns into a challenge.

Like health care images, medical signals likewise present quantity and speed snags, particularly during the persistent acquisition of high quality images and their storage from the many screens associated with every patient (Belle et al., 2015). Physiological signals not only create information dimension problems but also have baffling complexity of a spatiotemporal nature. Nowadays, numerous heterogeneous and uninterrupted monitoring devices are employed in the health care system to apply solitary physiological waveform information or crucial discrete data provided to systems if there should be an occurrence of plain occasion (Cvach, 2012; Drew et al., 2014).

The human genome is comprised of about thirty thousand genes. It has been observed that the price to sequence the human genome decreases with the advancement of high-throughput sequencing technology (E.S. Lander and et al., 2001; Drmanac et al., 2010). Investigating genome-scale information with suggestions for current public fitness insurance policies, the conveyance of care, and creating noteworthy proposals in an opportune way is a sizeable undertaking to the discipline of computational biology (Caulfield et al., 2013; Dewey et al., 2014). In a clinical setting, the delivery of these recommendations are very costly as time is very crucial.

In spite of huge expenditures by the current health care systems, clinical results become minimal (Oyelade et al., 2015). For big data analytics, it is a hazard to expect an extra quintessential section to help with investigation and revelation measures, bettering the conveyance of care, helping with the plan and design of medical care strategy, and make use of a way to exhaustively assess the muddled and tangled medical services information. More specifically, appropriation of the bits of information received from big data analytics can possibly save lives, enhance care conveyance, prolong admittance to clinical services, regulate installment to execution, and assist managing the improvement of clinical offering costs (Belle et al., 2015).

1.3 AI in Healthcare

Artificial intelligence impacts the health care domain as AI is the development of computer systems able to perform tasks that requires human intellect. Tasks such as object detection, decision making, solving complex problems, and so on are a few main benefits of artificial intelligence. AI also gives us predictions with an increased level of accuracy, it helps in decision making processes, it has solved complex problems, and it quickly performs high-level computations that take days for a human to solve. AI is something that makes human lives easier by performing high level computations and solving complex problems.

According to the PricewaterhouseCoopers (PwC) report, artificial intelligence will contribute an additional $ 15.7 trillion to the world economy by 2030, and the greatest impact will be in the health care sector (pwc). Healthcare is getting more import and using AI in more advanced manner. The sudden importance of AI in the health care industry can be categorized into two major points. First, the high availability of medical data; many of us have tons and tons of medical data in the form of our medical history and the availability of data makes implementing artificial intelligence much easier (Bush, 2018). Second, the introduction of complex algorithms. Machine learning alone is not capable of handling high dimensionality data and medical histories are extremely high dimensional in character, there are thousands of attributes that are hard for humans to analyze and process data through machine learning. However, when neural networks and deep learning were introduced, the process become much easier. Neural networks and deep learning are focused on solving complex problems that involve high dimensionality data; their development played a significant role in the impact of AI on health care (Simon et al., 2007; Loria).

AI benefits health care organizations by implementing cognitive technology to unwind a huge amount of medical records and perform power diagnosis. For instance, Nuance is a production service provider that uses artificial intelligence and machine learning to predict the intent of a particular user. By implementing Nuance in organization system to develop a personalized user experience, a company can make better actions that enhance customer experience and overall benefit the organization. Nuance helps in the storing, collecting, and reformatting of data to provide faster and more consistent access to allow further analysis or diagnosis. These are examples of how AI is gaining attention and being helpful to the health care industry (Aronson and Rehm, 2015; Schmidt-Erfurth et al., 2018).

1.4 Cloud Computing in Healthcare

There are numerous advantages and benefits of cloud computing in health care as it stores data on demand and reduces operational costs for health care providers such as hospitals and clinics (Rostrom and Teng, 2011). The cloud also supports electronic health records (EHRs), mobile applications that are used for medical concern applications. The cloud has made access to patient's medical history easier for doctors, and patients can now have a clear, in depth track of their medical process as well as track their appointments, results, and reports. Traditional EHRs need a team of medical staff, physicians, medical administrator, and IT members for their smooth processing and management; however, cloud EHRs do not require highly skilled IT members or professionals to develop and manage the system. Cloud EHRs make it easier to access data from anywhere than traditional EHRs system. Cloud computing works on either a one-time or monthly payment method and is less time consuming, whereas tradition EHR systems are costly and time-consuming. Cloud EHR systems also ensure that data are always kept safe in the cloud and can be retrieve when required (Mell and Grance, 2011).

Cloud Computing is significant in the health care sector as it supports big data analytics that are being used to improve decision support systems and contribute to therapeutic strategies in beneficial ways.

1.5 IoT in Healthcare

The internet of things is revolutionizing health care (Keh et al., 2014; Santos et al., 2014; Amendola et al., 2014). As patients become more connected and generate more data, clinicians can identify and address their needs more efficiently than ever before. And with advances in data science and artificial intelligence, the potential for personalized preventative care and other innovations is limitless in health care. In addition, IoT wearable devices help monitor patients' conditions with sensors for body temperature, heart rate, blood pressure, breathing, sleep, and much more that will help doctors or physicians know the medical and lifestyle history of an individual as well as detect early signs of critical diseases. The IoT saves time as well as money and when both things are saved, a life could be saved. For instance, if a person becomes erratic in the middle of night or is alone and experiences symptoms like shortness of breath, heart palpitations, and a sudden change in heart rate or blood pressure, the IoT wearable device will sense the rapid change, record it electronically, and transmit the data to medical providers. Back at the hospital, doctors could already be monitoring and evaluating the data.The nearest clinic or hospital could then provide an ambulance, and before the patient reaches the hospital, all the necessary equipment could be prepared for quick action. In this way, a doctor can save the life of a patient because the IoT was communicating with the health care system and by providing real-time monitoring to doctors, saved time, money, and effort. In this way, the IoT is transforming health care in a beneficial way (Lee, 2014; Yang et al., 2014; Diogo et al., 2014).

1.6 Conclusion

Smart health care informatics involve the combined use of the IoT, big data analytics,cloud computing, and artificial intelligence. These technologies will be made of use in health care by the application of artificial intelligence to examine and fit a giant quantity of data to screen for exclusion standards and decide the most appropriate objectives, keeping the time of recruiting the topics, and enhancing the concentrated efforts on the goal population. Further, patients are supervised in actual time with the usage of smart wearable devices to acquire extra time-sensitive and correct information, for instance, the practice of smart devices to display data in lung ailment clinical trials. Using big data analytics and synthetic genius in the field of health care research and the improvement of drugs will grow to be of greater convenience. The arrival of smart health care, mature standards, and structures has been organized. However, with the upcoming modern technologies, there is a huge scope for development, and many challenges are now coming out.

References

S. Amendola, R. Lodato, S. Manzari, C. Occhiuzzi, and G. Marrocco. RFID technology for IoT-based personal healthcare in smart spaces. *IEEE Internet of Things Journal*, 1(2):144–152, 2014. doi:10.1109/JIOT.2014.2313981.

S.J. Aronson and H.L. Rehm. Building the foundation for genomics in precision medicine. *Nature*, 526(7573):336–342, Oct 2015.

A. Belle, R. Thiagarajan, S.M.R. Soroushmehr, F. Navidi, D.A. Beard, and K. Najarian. Big data analytics in healthcare. *BioMed Research International*, 2015:370–194, Jul 2015. doi:10.1155/2015/370194.

J. Bush. How ai is taking the scut work out of health care, Mar 2018. URL https://hbr.org/2018/03/how-ai-is-taking-the-scut-work-out-of-health-care. Accessed: 2021-04-30.

T. Caulfield, J. Evans, A. McGuire, C. McCabe, T. Bubela, and et al. Reflections on the cost of "low-cost" whole genome sequencing: Framing the health policy debate. *PLOS Biology*, 11(11):1–6, 11 2013.

M. Cvach. Monitor Alarm Fatigue: An Integrative Review. *Biomedical Instrumentation & Technology*, 46(4):268–277, 07 2012.

M.E. DeBakey. The National Library of Medicine: Evolution of a Premier Information Center. *JAMA*, 266(9):1252–1258, 09 1991.

F.E. Dewey, M.E. Grove, and et al. Clinical Interpretation and Implications of Whole-Genome Sequencing. *JAMA*, 311(10):1035–1045, 03 2014.

P. Diogo, L.P. Reis, and N.V. Lopes. Internet of things: A system's architecture proposal. In *2014 9th Iberian Conference on Information Systems and Technologies (CISTI)*, pages 1–6, 2014. doi:10.1109/CISTI.2014.6877072.

B.J. Drew, P. Harris, J.K. Zègre-Hemsey, T. Mammone, D. Schindler, and et al. Insights into the problem of alarm fatigue with physiologic monitor devices: A comprehensive observational study of consecutive intensive care unit patients. *PLOS ONE*, 9(10):1–23, 10 2014.

R. Drmanac, A.B. Sparks, M.J. Callow, and et al. Human genome sequencing using unchained base reads on self-assembling dna nanoarrays. *Science*, 327 (5961):78–81, 2010.

S. Durcevic. 18 examples of big data analytics in healthcare that can save people, Oct 2020. URL https://www.datapine.com/blog/big-data-examples-in-healthcare/. Accessed: 2021-04-30.

B. Birren E.S. Lander, L.M. Linton and et al. Initial sequencing and analysis of the human genome. *Nature*, 409(6822):860–921, 2001.

H.C. Keh, C.C. Shih, K.Y. Chou, and et al. Integrating unified communications and internet of m-health things with micro wireless physiological sensors. *Journal of Applied Science and Engineering*, 17(3):319–328, January 2014.

B.M. Lee. Design requirements for iot healthcare model using an open iot platform. pages 69–72, 12 2014. doi:10.14257/astl.2014.66.17.

Keith Loria. Putting the ai in radiology. URL https://www.radiologytoday.net/archive/rt0118p10.shtml. (accessed 30 April 2021).

P. Mell and T. Grance. The nist definition of cloud computing. In *Computer Security Division, Information Technology Laboratory, National Institute of Standards and Technology*, pages 1–7, September 2011.

J. Oyelade, J. Soyemi, I. Isewon, and O. Obembe. Bioinformatics, healthcare informatics and analytics: an imperative for improved healthcare system. *International Journal of Applied Information System*, 13(5):1–6, 2015.

PricewaterhouseCoopers. Total economic impact of ai in the period to 2030. URL https://www.pwc.com/gx/en/issues/data-and-analytics/publications/artificial-intelligence-study.html. Accessed: 2021-04-30.

F. Ritter, T. Boskamp, A. Homeyer, H. Laue, M. Schwier, F. Link, and H.O. Peitgen. Medical image analysis. *IEEE Pulse*, 2(6):60–70, 2011. doi:10.1109/MPUL.2011.942929.

T. Rostrom and C.C. Teng. Secure communications for pacs in a cloud environment. In *2011 Annual International Conference of the IEEE Engineering in Medicine and Biology Society*, pages 8219–8222, 2011. doi:10.1109/IEMBS.2011.6092027.

A. Santos, J. Macedo, A. Costa, and M.J. Nicolau. Internet of things and smart objects for m-health monitoring and control. *Procedia Technology*, 16: 1351–1360, 2014.

U. Schmidt-Erfurth, H. Bogunovic, A. Sadeghipour, and et al. Machine learning to analyze the prognostic value of current imaging biomarkers in neovascular age-related macular degeneration. *Ophthalmology Retina*, 2(1):24–30, 2018.

J.A. Seibert. *Modalities and Data Acquisition*, pages 49–66. Springer New York, New York, NY, 2010. doi:10.1007/978-1-4419-0485-0_4.

D. Simon, A. Loh, and M. Härter. Measuring (shared) decision-making – a review of psychometric instruments. *Zeitschrift für ärztliche Fortbildung und Qualität im Gesundheitswesen - German Journal for Quality in Health Care*, 101(4):259–267, 2007.

G. Yang, L. Xie, M. Mäntysalo, X. Zhou, and et al. A health-iot platform based on the integration of intelligent packaging, unobtrusive bio-sensor, and intelligent medicine box. *IEEE Transactions on Industrial Informatics*, 10(4): 2180–2191, 2014. doi:10.1109/TII.2014.2307795.

2

A Review on Deep Learning Method for Lung Cancer Stage Classification Using PET-CT

Kaushik Pratim Das[1,*], *Chandra J*[1], *and Dr Nachamai M*[2]

[1] *Department of Computer Science, CHRIST (Deemed to be University), Postcode-560029, Hosur Road, Bangalore, India*
[2] *Siemens Healthcare Pvt. Ltd, Electronic City, Phase II, Postcode-5600100, Bangalore, Karnataka, India*

2.1 Introduction

The lung is a complex organ with several types of tissue. Given the nature of the lungs' complexity, dilemmas arise in detecting lung cancer at a preliminary stage. Early detection, accuracy in identifying the tumor, lung cancer stage, and the disease's prognosis are critical elements monitored during cancer staging (Huang et al., 2019). Based on observations during the staging procedure, the treatment plan is determined. Thus, cancer staging requires multiple alterations to accurately detect the tumor, which is challenging to achieve with rigid mechanisms and imaging techniques. Under circumstances that require curative surgery, the most strenuous process is the ability to extract the organ's features. Therefore, it is essential to separate these features from the tumor region with precision and without loss of detail for accurate lung treatment-based decisions.

In recent years, technical advancement has led to the adoption of multi-modalities for medical imaging. Lung cancer is an imaging-dependent sphere within the specialty of medicine. Positron emission tomography (PET) is considered an important imaging modality for diagnostic purposes and staging. As malignancy develops, a PET scan can detect abnormalities that may not be part of the malignancy. Although a PET scan is considered one of the efficient imaging modalities in medical imaging, the PET scan's sensitivity and specificity are uncertain.

Further advances have led to the fusion of modalities such as positron emission tomography–computed tomography (PET-CT) for the co-registration of medical images to achieve better imaging functionality and detection accuracy. In clinical practice, the standard procedure of cancer staging is called TNM staging (Brierley et al., 2016). Due to various internal and external factors related to lung cancer patients, cancer staging is not a simple task. As cancer has become complicated, the early detection and prognosis of cancer have become a necessity.

*Corresponding Author: Kaushik Pratim Das
email: kaushik.das@res.christuniversity.in

Applied Smart Health Care Informatics: A Computational Intelligence Perspective, First Edition.
Edited by Sourav De, Rik Das, Siddhartha Bhattacharyya, and Ujjwal Maulik.

Image processing has been the primary method for an efficient staging process that lead to successful lung cancer treatment. Identifying the anatomical positioning of a tumor, lung features, separation of healthy tissues from the tumor, and tumor classification are the foremost medical image processing tasks. Several types of research have implemented traditional approaches for image processing for lung cancer detection. Although the results are satisfactory, the traditional image processing models have a lesser scope of improvement than deep learning technologies. The advanced approaches enable the network to improve with training, and the outcome of this procedure allows the model to produce detailed results.

Deep learning technology is widely gaining acceptance across multiple disciplines such as medicine, biomedical engineering, and information technology. With automation and faster computation capabilities, deep learning approaches are implemented for an efficient image processing procedure with precision and accuracy in tumor detection. Incorporating deep learning into cancer imaging can facilitate efficient segmentation and classification of the lungs. Segmentation helps to identify the objects and boundaries and remove unwanted features from a medical image, which aids in minimizing the occurrence of treatment failure. The extent of manual detection is demanding for complicated cancer types, and advanced deep learning models have shown promising results in the segmentation of lung nodules using a Convolutional Neural Network (CNN) with an average accuracy rate of 80% (Wendy et al., 2019). The classification techniques that use deep learning have achieved tremendous success, allowing classification of the patients according to the cancer type, and the prognosis is another relevant contribution of image processing. Classifiers such as random forest, support vector machines (SVMs), k-nearest neighbors (KNNs), and decision trees have been the most efficient for classification tasks in a typical image processing procedure. With the rapid evolution of technology, deep learning-based techniques will garner more acceptability and become a tool for efficient staging and better clinical decision making in oncology.

2.1.1 Scope of the Research

The early detection of lung cancer is low. Therefore, numerous cases stay undetected until it has reached an advanced stage. Due to fatality factors associated with lung cancer, the treatment planning must be appropriate with an accurate diagnosis. Cancer staging helps eliminate any possibility of misdiagnosis as each stage of cancer is considered for diagnosis. Based on the stage of cancer, the suggested treatment begins for a patient. However, the lung's complex nature and the influence of internal and external factors cause the lung tumor to be challenging to detect. These additional factors cause the diagnostic image quality to deteriorate, making diagnosis and treatment complicated and time-consuming. In addition, there might be multiple diagnostic sessions for complex cases to have an accurate outcome.

With the advent of nuclear medicine, the diagnosis has significantly improved. However, the quality of images plays a vital role in nuclear medicine. High-quality images ensure that accurate diagnosis occurs regularly. The current medical imaging technologies lack automation in segmentation and detection procedures. Manual detection of the cases becomes laborious and problematic when there are several patients in a clinical setting. Although PET scans are preferred for lung cancer detection, some challenges include high or low contrast in the images, blurry images, noise, and an inability to capture the metastasis. The accurate diagnosis of metastatic spread ensures that the appropriate region of interest is identified during detection, and treatment is suggested as per the acquired results. These challenges necessitate an accurate and automated image processing technology that can aid in faster diagnosis. In addition, the possibility of producing reliable results and image classification capabilities of the stage of cancer is crucial for further treatment.

This research aims to identify an efficient deep learning technology that may aid in the lung cancer staging procedure. Deep learning's potential makes it an effective solution for automation and classification of cancer stages from multiple images in less time while delivering reliable clinical information for accurate treatment planning. The proposed research's primary motivation is to address the challenges of early lung cancer detection and its stage for accurate lung cancer patient diagnosis.

2.1.2 TNM Staging

Lung cancer is a significant cause of cancer-based mortality worldwide. The formulation of efficient treatment planning for the prognosis is highly dependent on an effective staging process. In the field of medicine, lung cancer staging is constructed with the TNM staging system. Under TNM staging, the disease is identified as a primary tumor (T), lymph node (N), and metastasis (M). The TNM staging is revised and based upon the eighth edition proposed by the International Association for the Study of Lung Cancer (IASLC) (Goldstraw et al., 2016). The primary focus on lung cancer staging is to verify its anatomical spread in a patient.

2.1.2.1 TNM Descriptors for Staging per IASLC Guidelines

The IASLC guidelines for TNM-8 staging have introduced changes from the previous edition. The TNM classification of lung cancer for staging is one of the most consistent indicators of disease prognosis (Kirienko et al., 2018). However, the classification does not provide the entire prognosis as lung cancer is affected by multiple factors. The prognosis of the disease and its recurrence after the treatment changes the type of further treatments at several phases. TNM staging does not include supplementary information related to the disease, such as analyzing the patient blood samples, markers used for tumors, and specific genetic attributes (Goldstraw et al., 2016). TNM staging provides a predictive analysis of lung cancer staging with each of the descriptors. However, there are limitations and clinical situations that have not been clarified in the staging by imaging procedure (Feng and Yang., 2019). As technology advances, the need for refinement in the staging process is a necessity. There is a constant need to eliminate some of the limitations and introduce automation to overcome the TNM staging complexities. Thus, deep learning can play a crucial role in TNM classification for staging.

2.1.2.2 PET-CT Scan in Lung Cancer Imaging

The past decade has witnessed CT scans as the gold standard among imaging methods. CT scans lack efficiency in differentiating between benign and malignant tumors. The PET scan was developed due to the observation that malignant cells were associated with an increased glycolic rate and an increased cellular glucose uptake. Using PET-CT scans has been widely considered for lung tumor detection. Images acquired with the use of PET-CT scans are helpful for the evaluation of lung cancer cases. PET-CT scan images are critical for the evaluation of pulmonary nodules that are not identified by a biopsy. The fused images of both modalities are also useful for cases that involve abnormalities. A PET-CT scan is considered essential in complicated cases that may result in the clinician's hesitation for curative surgery due to an indefinite diagnosis of a suspected lesion. PET scans are efficient, and possess the capability for accurate tumor localization. PET-CT scans provide more indications on the pulmonary nodules' morphological and functional characteristics and is used for TNM staging where a CT scan alone cannot determine the metastasis. Based on cost factors, the PET-CT fusion has proven to be cost-effective as the fusion can occur in a single system and be analyzed side-by-side. PET scan procedures are lengthy, and the image fusion of PET and CT images can save up to 20 to 30 minutes per patient (Saif et al., 2010). Although many different forms of advanced imaging modalities have been developed in recent years, such as MRI, PET-MRI, 4DCT, 4DMRI, they are costly, and most of the advanced modalities barring MRI have yet to be introduced in developing nations.

PET images help attain the functional information while CT images provide the anatomical information, thereby facilitating complete information on a malignancy (Teramoto et al., 2019). Besides, lung cancer patient restaging determines a patient's response after treatment or surgery and provides more insight into the lesion histology. Clinical advances have made PET-CT a useful indicator for the prognosis of the disease. Integrating PET-CT produces more accurate TNM classification characterization results than PET or CT individually (Pinilla et al., 2008). The lungs are exposed to many external and internal challenges regularly. Therefore, when a malignancy develops, the PET scan is susceptible to detecting abnormalities that are not part of the malignancy. PET scans are not considered among the best in terms of sensitivity and specificity, so integration with CT scans is preferred in current imaging modalities. PET-CT images are prone to misregistration, which could lead to misdiagnosis (Hochhegger et al., 2015). The limitations of existing diagnostic modalities have shifted the focus to embracing efficient technological aids to improve the overall accuracy of diagnosis and treatment. Consequently, the acceptance of technical aid has led to immence contributions by artificial intelligence.

2.2 Related Works

Wang et al. (2019) designed a three Deep Convolutional Neural Network (DCNN) for lung cancer detection. The proposed model used co-learning from chest CT scan data to identify benign and malignant tumor regions without prior anatomical delineation of the suspected nodule. XGBoost was used as a classifier for the prediction outcome. The performance accuracy was measured as 78% for the Area Under the Curve (AUC).

Song et al. (2017) proposed a deep neural network to classify benign and malignant pulmonary nodules. The research involved a performance comparison for deep neural networks (DNN), CNNs, and stacked autoencoders (SAE). The models were evaluated on the The Lung Image Database Consortium (LIDC-IDRI) database. The study results showed that CNN achieved a better performance with an accuracy of 84.15%, sensitivity of 83.96%, and and specificity of 84.32%.

Bhatia and Sinha (2018) proposed a deep residual learning approach for detecting lung cancer from CT scans. The research includes a pipeline comprising of preprocessing techniques to delineate cancerous lung regions. The features were extracted using UNet and ResNet computational models. Classifiers such as XGBoost and random forest were fed with features to predict lung cancer's likelihood from the data set. The performance accuracy was 84%.

Yiwen et al. (2019) evaluated a deep learning network to predict lung cancer treatment outcomes using CT images. The patients with an advanced stage of non-small cell lung cancer carcinoma were considered. The pre-treatment and post-treatment results were taken into consideration at one, three, and six months follow-up. Transfer learning was used for the DCNN. A recurrent neural network (RNN) was used to analyze the survival predictions and cancer-related outcomes. The accuracy was found to be 74% using the AUC.

Deng et al. (2020) proposed a feature fusion approach based on PET images and additional clinical information. The purpose of the research was to obtain the features for predicting the lung metastasis of soft tissue sarcomas. Random forest was used to eliminate redundant features. The proposed method achieved a satisfactory performance when fusion features were used along with three standard uptake values. The accuracy was found to be 92%.

Weikert et al. (2018) evaluated an artificial intelligence-based lung nodule algorithm for detection and segmentation of primary lung tumors that covered all T-staging categories. PET-CT was used for the experimentation. It was noted that current applications for lung nodules exist to detecting those less than 3 cm in size. The results showed that the detection accuracy was best in the T1 category with 90.4% and significantly decreased to 70.8% for T2, 29.4% for T3, and 8.8% for T4 stages.

Wu et al. (2012) evaluated the Ensemble Algorithm of cancer data clustering that groups patients according to the prognostic factors for TNM staging. The research demonstrated that the algorithm generates results that are in line with the AJCC staging approach, which could provide a mechanism to add other prognostic factors for a complete estimation of the prognosis and survival.

Kourou et al. (2015) presented a review on machine learning application for cancer prognosis and prediction. It was stated that several techniques such as Artificial Neural Network (ANN), Bayesian networks, decision trees, and SVMs are widely used for cancer research that targets accurate decision making regarding malignancies. Although machine learning can improve cancer progression, an appropriate validation is needed for use in clinical practices. The addition of more diversified data combined with techniques for feature selection and classification could prove a promising tool for cancer research.

Kirienko et al. (2018) proposed a deep learning algorithm based on CNN for lung cancer assessment with T-staging parameters that used PET-CT data. The 7th edition of TNM staging was used as a reference. The CNN's input was in the form of a bounding box on both PET and CT images. The classified results were incorrect. The results were reported as accuracy, recall, and specificity, while the algorithm comprised two networks: a feature extractor and a classifier. The final model's accuracy was 86.8% for training, validation accuracy was 69.3%, and test accuracy was 69.1%.

Qin et al. (2020) presented a deep learning architecture that combines fine-grained features from PET and CT images. The study focused on featured noise reduction with a multidimensional mechanism while extracting the fine-grained features for both the imaging modalities. The AUC was used for measuring the performance. The average score was 92% (0.920) with a variance of 0.05 and a balanced accuracy of 72% (0.72).

2.2.1 Artificial Intelligence in Medical Imaging

The emergence of AI has been followed by a rapid growth in its integration in medical imaging. Factors such as automation, less computational time, and high-quality images are the foremost considerations of adopting new technology, especially in a high-risk environment such as cancer diagnosis. The workloads of radiologists and oncologists have grown significantly over the years, with the number of cases increasing in recent years. Studies have reported that a radiologist elucidates an image every three to four seconds (Hosny et al., 2018). Considering the rate per second for interpretation of the image, it becomes inevitable to be erroneous as a radiologist and an oncologist in the diagnosis and prognosis of patient disease.

With the significance of staging lung cancer and the limitations of imaging modalities, there is a need for a seamless automation component in the cancer imaging workflow. The requirement to refine the staging process with the help of technologies based on artificial intelligence can enhance tumor classification efficiency in cancer staging. The accurate classification results are interlinked with the successful completion of treatment, thus minimizing the errors and treatment failures by assisting the oncologist with well-defined images of the identified Region of Interest (ROI).

2.2.2 Classification for Medical Imaging

Classification techniques are performed to classify the data into various classes. The data is arranged and categorized for efficient use. The contribution of deep learning techniques for image classification in medical imaging has made significant contributions. There are multiple image modalities, e.g. CT, PET, and MRI for cancer diagnosis. However, several of these procedures are expensive and time-consuming. In recent years, various studies have indicated that lung cancer detection has occurred at advanced stages, which proves to be fatal for patients. Often, advanced stage cases have a lower survival rate. The contribution of AI has been targeted to help diagnose lung cancer with an eye on the cost-effectiveness of treatment. Image-processing techniques such as classification are significant contributors to a successful diagnosis and prognosis of the disease. Jingchen et al. developed a computer-aided diagnostic (CAD) method to classify the lung nodule that incorporated radiomics and applied random forest. The proposed model could produce higher accuracy in terms of sensitivity and specificity, and it can be applied to a large data set, as noted by the authors. The sensitivity was found to be 80%, and the specificity was 85.5%.

2.2.2.1 Deep Learning

Deep learning is a sub-field of ML. In recent years, deep learning algorithms have gained a considerable amount of acceptance in medical imaging. The capability of deep learning algorithms to automatically learn the representation of regions and features from the data without human intervention has been the core of these techniques. Deep learning-based models have extracted quantifiable features with a high accuracy for subtracting the human tissues from the region of interest that promises improvements in tumor diagnosis. There are substantial benefits of eliminating manual supervision using deep learning, such as preprocessing and segmentation, which is a valuable addition for efficient clinical decisions based on prognosis and treatment recommendations.

2.2.2.2 Image Classification Using Deep-learning Techniques

A clinical setting where each diagnosis constitutes a sample, leads to generation of a large data set. Transfer learning has played a vital role in image classification for processing such large data sets. A pretrained network is used on a large data set to train all parameters of the neural network. Two strategies are used for transfer learning: (i) using a pretrained network that works as the feature extractor and (ii) fine-tuning the pretrained network for the medical image data set. Using the former strategy as a feature extractor has the benefits of not requiring to train the deep network while allowing for the use of extracted information for image analysis in the pipelines (Litgens et al., 2017). A highly successful result based on the pretrained network as the feature extractor was achieved by Esteva et al. (2017).

2.3 Methods

2.3.1 Transfer Learning

According to the data set's size, a deep learning model's training can take several hours, days, or weeks. The processing power required for the training is high and is a critical requirement for the workflow. However, for large data sets, a Graphics Processing Unit (GPU)-enabled computation is preferable for speeding up the workflow. Thus, transfer learning is used to retrain the model for the targeted problem (Paul et al., 2016). Transfer learning allows for application of the knowledge of a particular problem to the targeted problem of similar criteria. For successful transfer learning, the last three layers of the pretrained network are modified and fine-tuned to retrain the model with the required data set. To perform transfer learning, three components are of utmost importance.

1. The set of layers represents the network architecture. For transfer learning, the layers are created by modifying a pre-existing network for AlexNet.
2. The images with appropriate labels act as a data store to train the model.
3. The use of a variable to control the behavior of the training algorithm. Such components are used as the inputs to the `trainNetwork` function that returns the trained network as output.

2.3.2 AlexNet

AlexNet has been considered one of the most useful CNN architectures. AlexNet consists of a large data set with millions of images with high-resolution features and 1000 categories. With 60 million parameters and 650 000 connections, faster processing is achievable with GPU and non-saturating neurons. However, a dropout method in the AlexNet architecture helps to avoid overfitting problems. AlexNet consists of eight layers: three convolutional layers, two pooling layers, and two fully connected layers with a SoftMax classifier. The network's output layer can provide 1000-way SoftMax that can recognize 1000 class labels. The overfitting problem is reduced in the fully-connected layers with a method called dropout. Fine-tuning a network using transfer learning allows for faster training of a network as learned features can be transferred to perform a new task using a smaller set of training images (Krizhevsky et al., 2012). AlexNet is proficient in classification problems and performs best on a larger data set. However, the performance is likely to reduce if any convolutional layer is missing in the network.

2.3.3 AlexNet Architecture

Figure 2.1 depicts the AlexNet architecture for this research where the input layer accepts images in a network that is sized at 227 x 227 x 3. The input will have three channels if the input is a color image. The procedure of the network is described as follows.

1. Convolutional Layer 1: In the first convolutional layer, the filter size is $f = 11$, and the number of filters is 96. The stride of four is used, which helps decrease the dimensions of an input volume by four. The output after the first convolution is 55 x 55 x 96 volume. The mathematical representation of a convolution is defined as the product of "f' and 'g " object functions. The two functions f and g are over the range of $[0, t]$, where [f*g] (t) denotes the convolution of f and g. Max pooling is one, as given in (equation 1) (Sajja et al., 2019).
2. Maxpooling 1: This layer comprises a 3 x 3 filter and a stride of two. The reduction in 55 x 55 x 96 volume to 27 x 27 x 256 is due to the strides use in this layer.
3. Convolutional Layer 2: The same convolution is used in this layer with a filter size of f = 5, which produces the same 27 x 27 x 256 dimensions.

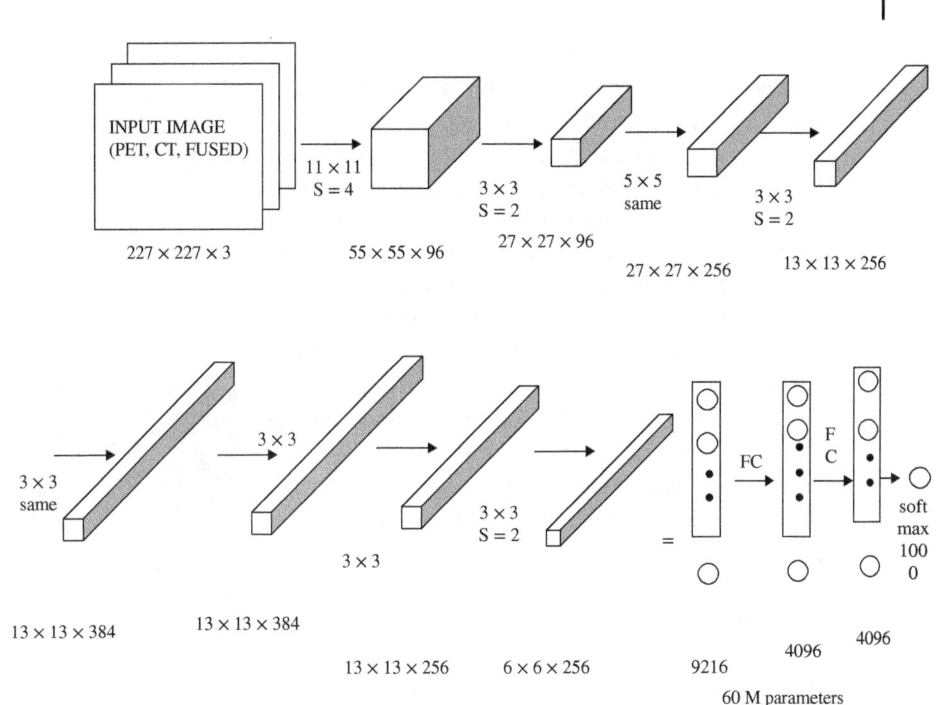

Figure 2.1 AlexNet architecture.

4. Maxpooling Layer 2: The height and the width are reduced to 13 in this layer using a 3 x 3 filter, and stride two is used.
5. Convolutional Layers 3, 4, and 5: The 3 x 3 convolution layers with padding = 1 and a stride = 1 is being applied. In the first two convolutional layers, 384 filters are used, whereas, in the Conv5 layer, 256 filters are used.
6. Maxpool3: This layer has a stride of 2 such that the volume with dimensions 6 x 6 x 256 is achieved. The multiplication of the numbers results in 9216, which are passed into 9216 nodes.
7. Dropout: The dropout technique has a setting of zero for the hidden neuron's output of a probability of 0.5. The neurons which fall under dropout do not contribute to the forward propagation or the backward propagation process. In the input event, the neural network processes a different architecture while sharing the weights among them. This technique reduces the need for coadaptations, thus allowing the network to learn more robust features.
8. AlexNet comprises three fully connected layers. The first two layers comprise 4096 nodes, whereas the third fully connected layer has 1000 units. SoftMax is used to produce the output for classification.

2.3.4 Experimental Setup

The images were collected from "Imaging of lung cancer: Implications on staging and management" by Purandare (2015), "PET-CT in staging lung carcinoma" by

Kochhar (2019), and the Lung PET-CT Dx data set from the *Cancer Imaging Archive* (Angelus and Kirby, 2020). Figures 2.2 and 2.3 are medical images of lung cancer stages. For experimentation, the source's images were sorted as per the TNM staging of lung cancer and compiled into a small data set of 101 images. The purpose of having a diversified data set is that there will be several types of medical complexity for a lung cancer patient during the staging of cancer, and the stages of cancer will be varying. There are also the subproblems of an underlying disease that needs to be addressed. Therefore, the aim was to cover all the possible underlying factors during staging and to label the data set according to the diseases. The AlexNet was fine-tuned for this problem with a small data set, as the extent of the diseases during staging will generate a massive data set that might extend beyond the research scope. The data set included annotations provided by the radiologist. The annotations included TNM staging information, type of lung cancer, tumor diameter, and the disease's related subcriteria. Additionally, the annotations included PET-CT acquisition and correction information from the fusion of PET-CT images during staging and treatment change.

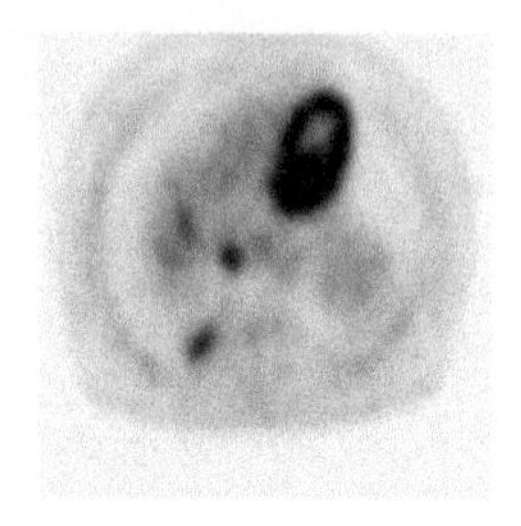

Figure 2.2 Metastasis-PET image.

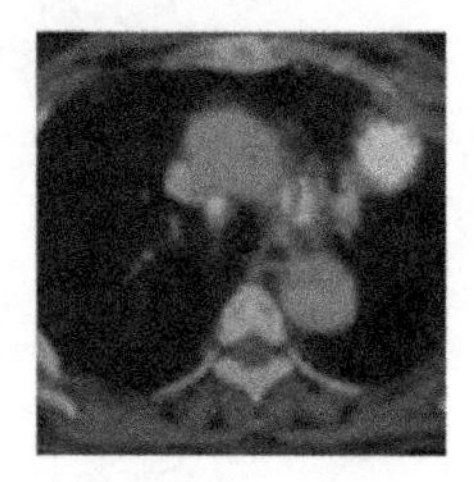

Figure 2.3 Lymph node-fused PET-CT image.

Table 2.1 depicts the training setup of the AlexNet that was implemented with a transfer learning approach and fine-tuning the network for research. The epoch was 20, as the data set is not large. The epoch for the training was kept at a minimum due to the use of transfer learning for the training. The maximum iterations during the training process were 100 for the primary tumor, 40 for the metastasis, and the lymph node had 20 iterations.

2.3.4.1 Image Processing

The tool used for experimentation is MATLAB. The medical images were loaded using an image data store that automatically labels the images based on the folder names, and the data is stored as an `ImageDatastore` object. The image data store can store large image data and efficiently read images in batches during the network's training. The data set is further divided into training and validation sets,

Table 2.1 Training setup for AlexNet.

Lung Cancer Stages	Max. Epoch	Max. Iterations	Iterations/ Epoch
Primary Tumor	20	100	5
Metastasis	20	40	2
Lymph Node	20	20	1

where 70% of images were used for training and 30% for validation. The data set has three classes: (i) primary tumor (T) with the primary subclasses Stage(T1), Stage(T2), Stage(T3), and Stage(T4); (ii) metastatic (M) with the subclasses metastasis M1a, metastasis rib, metastatic adenoma, metastatic nodes, and metastatic pleural deposits; and (iii) lymph node (N) with the subclasses nodal disease, and Stage(N3).

For the new classification task, the last three layers of the network were fine-tuned and replaced with a fully connected layer, SoftMax layer, and a classification output layer. The fully connected layer is modified according to the new data, whereas the layer's size is similar to the number of classes in the data set. The `WeightLearnRateFactor` and `BiasLearnRateFactor` were applied to achieve faster learning of the new layers in the network.

2.3.4.2 Data Augmentation

The input layer of the AlexNet accepts data with an image size of 227 x 227 x 3, where the three denotes the number of color channels. The augmented image data store was used to resize the training images and perform basic preprocessing tasks automatically. The overfitting problem was prevented using data augmentation and by enabling the network to memorize the exact details of training images.

2.3.4.3 Training and Validation

For the implementation of transfer learning, the features from the layers of the pre-trained network were retained. The transferred layers had an initial learning rate of a smaller value to slow the learning compared to the fully connected layer in which the learning rate was increased to speed up the learning of the new layers. A stochastic gradient descent algorithm (SGDM) was implemented to update the learnable parameters for network training. SGDM also enabled minimizing the loss function. The training of the network was performed on an NVIDIA-enabled central processing unit (CPU). The validation images were classified using the fine-tuned AlexNet and displayed with the predicted labels. The accuracy of the validation set was calculated, which provided the results of the correct label prediction by the network. A confusion matrix was plotted to display the results of the output class and the targeted class. The maximum epoch size of 20 was used to perform data training. The results obtained are highlighted in the following figures.

2.4 Results and Discussion

2.4.1 Primary Tumor (T)

Figures 2.4–2.7 depict results of the primary tumor, the T stage of TNM staging, which consists of subclasses for the stage of cancer including T1, T2, T3, and T4. The cancer stage begins at stage 1 and progresses to stage 4 with multiple additional possibilities such as adenocarcinoma, curative or non-curative, and satellite nodules.

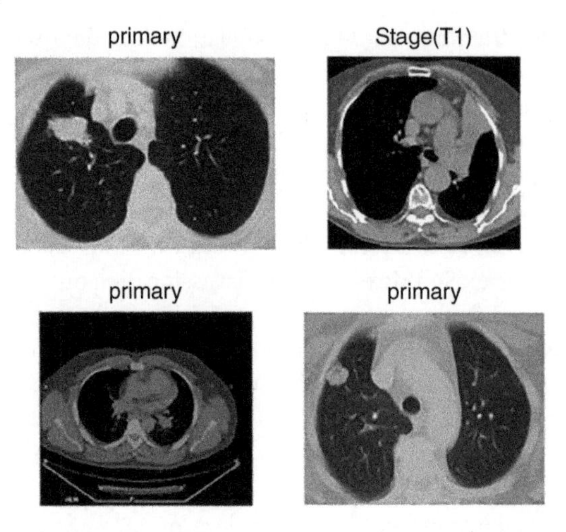

Figure 2.4 Classification results of the primary tumor (T).

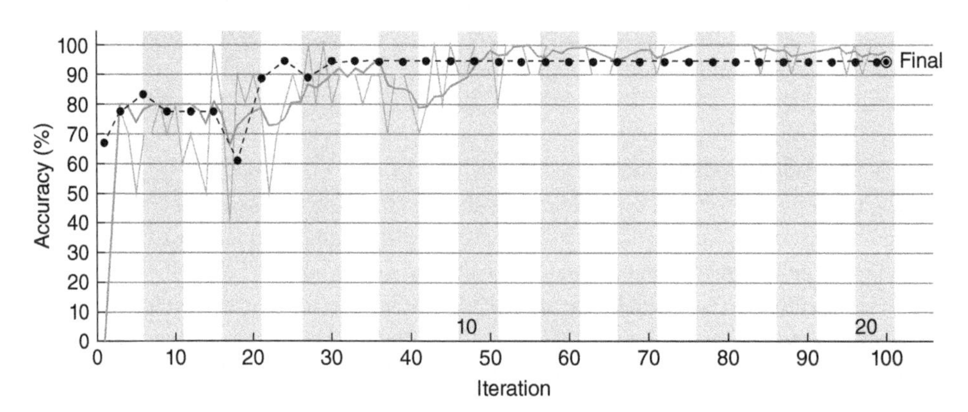

Figure 2.5 Classification accuracy of the primary tumor (T).

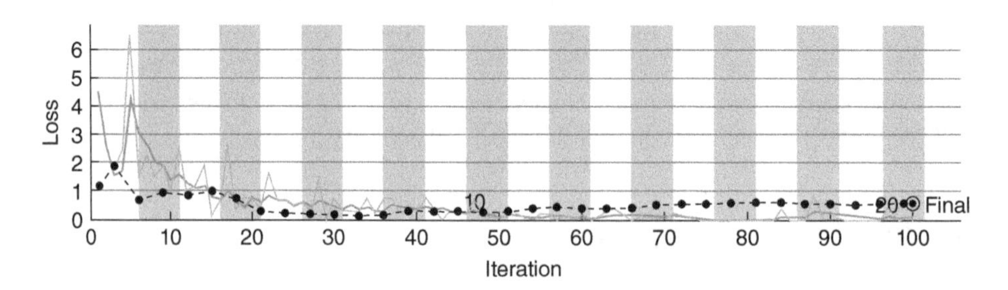

Figure 2.6 Loss function for the primary tumor (T).

The validation accuracy of the classification task was 94.44%. Figures 2.6 and 2.7 depict the loss function and the confusion matrix, which is plotted for the results of the output class and the targeted class to show the performance of the classification accuracy of the fine-tuned AlexNet.

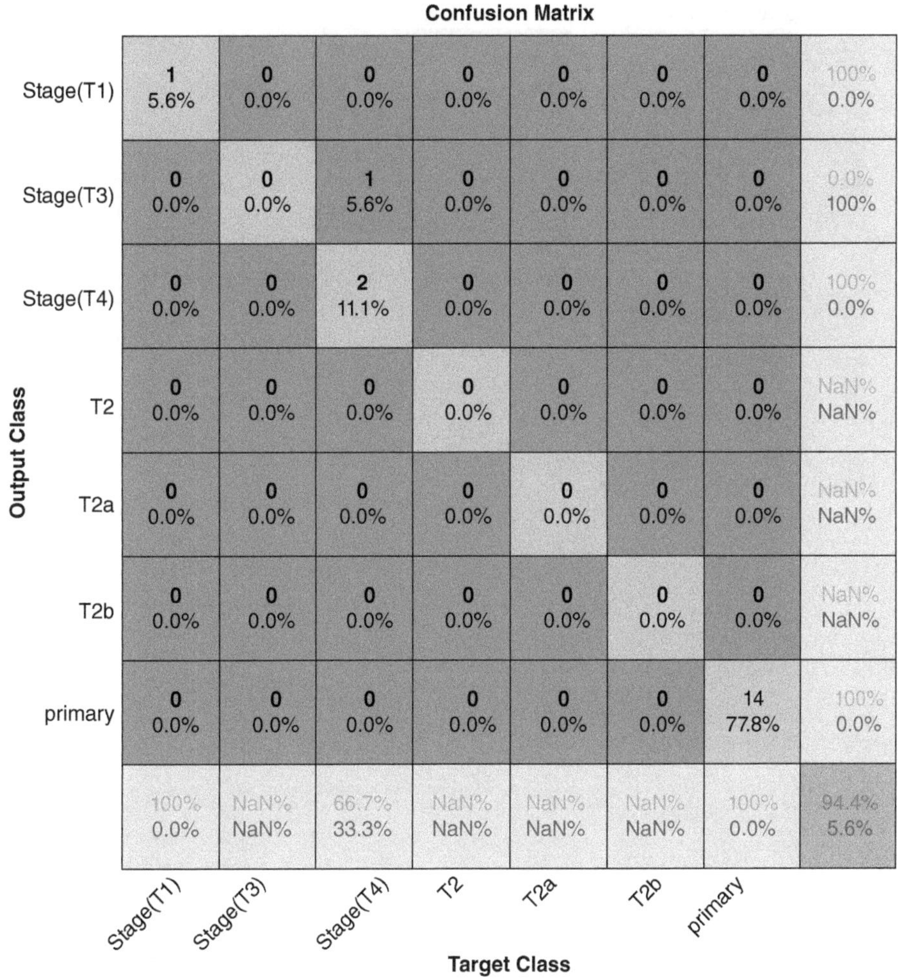

Confusion Matrix

Output Class \ Target Class	Stage(T1)	Stage(T3)	Stage(T4)	T2	T2a	T2b	primary	
Stage(T1)	1 5.6%	0 0.0%	0 0.0%	0 0.0%	0 0.0%	0 0.0%	0 0.0%	100% 0.0%
Stage(T3)	0 0.0%	0 0.0%	1 5.6%	0 0.0%	0 0.0%	0 0.0%	0 0.0%	0.0% 100%
Stage(T4)	0 0.0%	0 0.0%	2 11.1%	0 0.0%	0 0.0%	0 0.0%	0 0.0%	100% 0.0%
T2	0 0.0%	0 0.0%	0 0.0%	0 0.0%	0 0.0%	0 0.0%	0 0.0%	NaN% NaN%
T2a	0 0.0%	0 0.0%	0 0.0%	0 0.0%	0 0.0%	0 0.0%	0 0.0%	NaN% NaN%
T2b	0 0.0%	0 0.0%	0 0.0%	0 0.0%	0 0.0%	0 0.0%	0 0.0%	NaN% NaN%
primary	0 0.0%	0 0.0%	0 0.0%	0 0.0%	0 0.0%	0 0.0%	14 77.8%	100% 0.0%
	100% 0.0%	NaN% NaN%	66.7% 33.3%	NaN% NaN%	NaN% NaN%	NaN% NaN%	100% 0.0%	94.4% 5.6%

Figure 2.7 Confusion matrix for the primary tumor (T).

2.4.2 Metastasis (M)

Figures 2.8–2.11 depict the accuracy of classification of the metastasis class. The subclasses under metastasis are related to the various forms of metastasis that can occur when cancer has spread from the lungs to other parts of the body. The subclasses consist of the metastatic node, M1a stage of metastasis, metastatic adenoma, metastasis rib, and metastatic pleural deposit. The validation accuracy for classification was 100%. Figures 2.9–2.11 represent the training progress, loss function, and the confusion matrix, respectively, for evaluating the performance of the network's classification accuracy.

2.4.3 Lymph Node (N)

Figures 2.12–2.15 show the classification accuracy of the network for the lymph node class. The subclasses that were added are Stage(N3) and nodal diseases. The

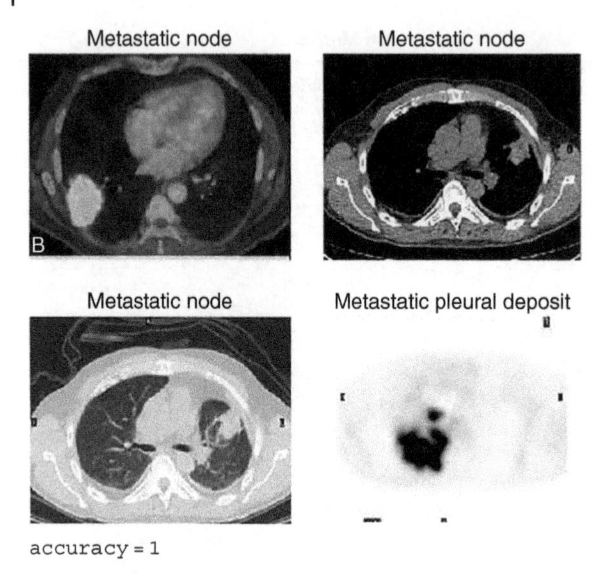

Figure 2.8 Classification results of the metastasis (M).

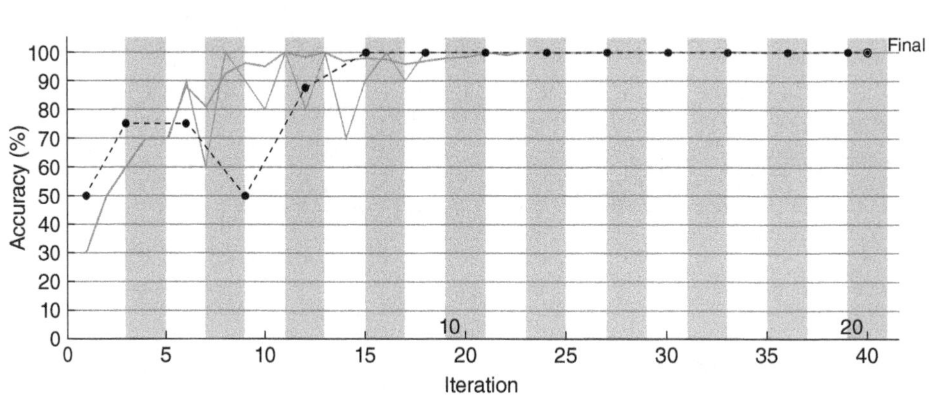

Figure 2.9 Classification accuracy of metastasis (M).

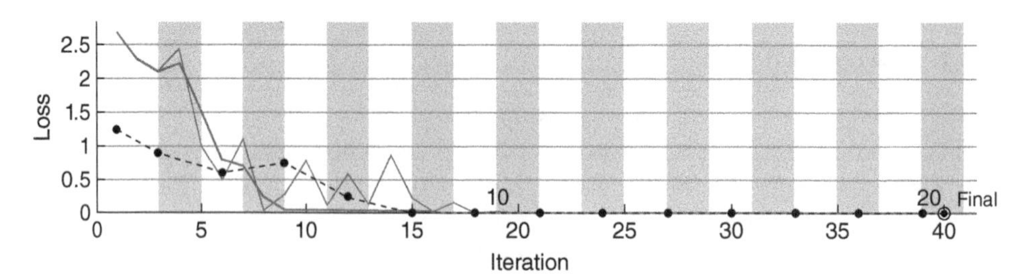

Figure 2.10 Loss function for the metastasis (M).

network achieved a validation accuracy of 100% for the classification of the data set in this class. Figures 2.13 and 2.14 represent the training results and the loss function, respectively. Figure 2.15 depicts the confusion matrix plot for evaluating the network's classification performance.

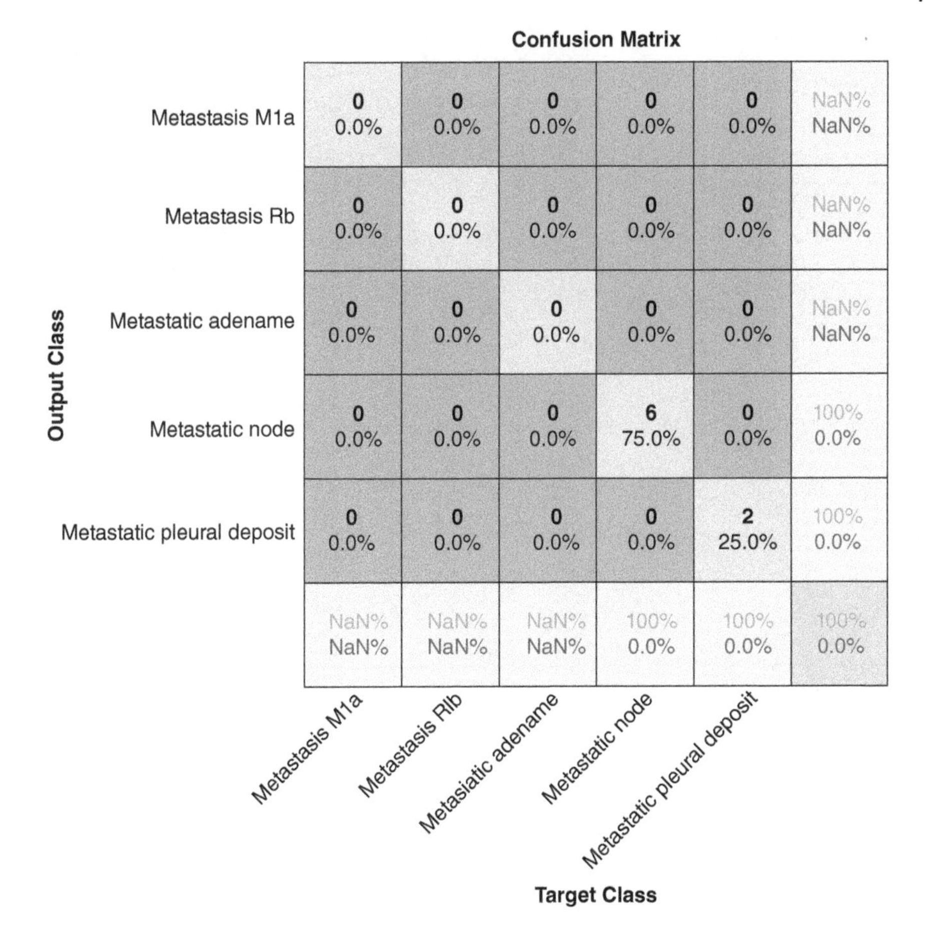

Figure 2.11 Confusion matrix for the metastasis (M).

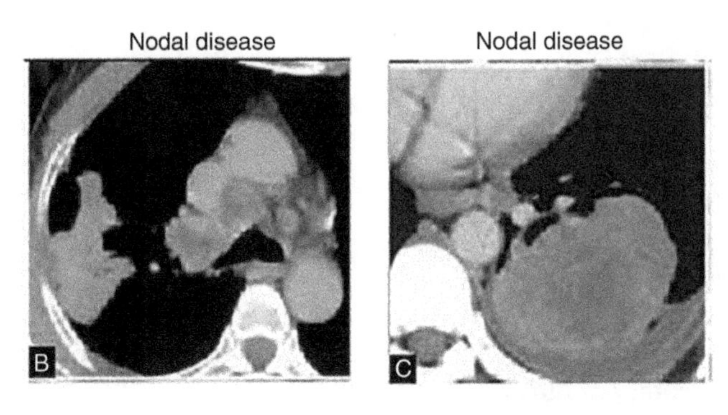

`accuracy = 1`

Figure 2.12 Classification results of the lymph node (N).

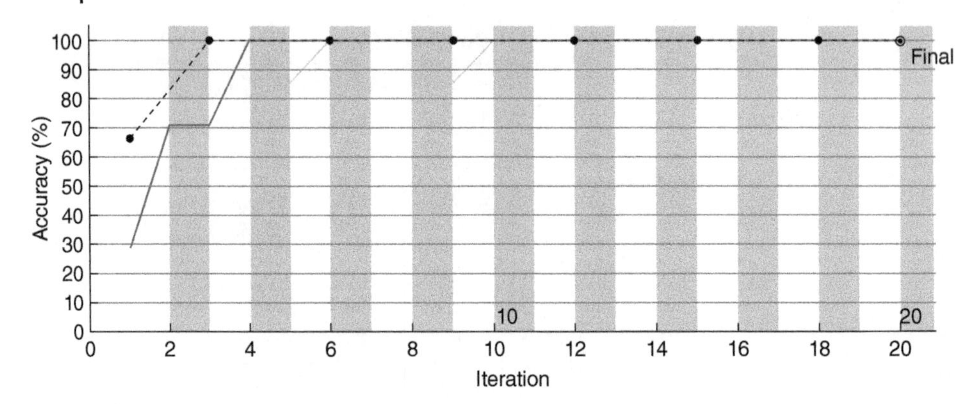

Figure 2.13 Classification accuracy of the lymph node (N).

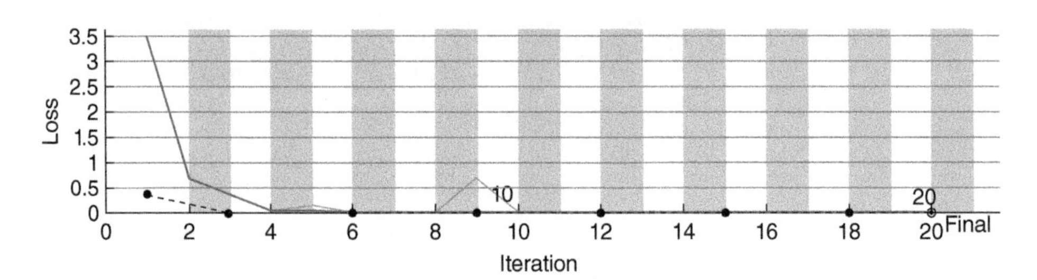

Figure 2.14 Loss function for the lymph node (N).

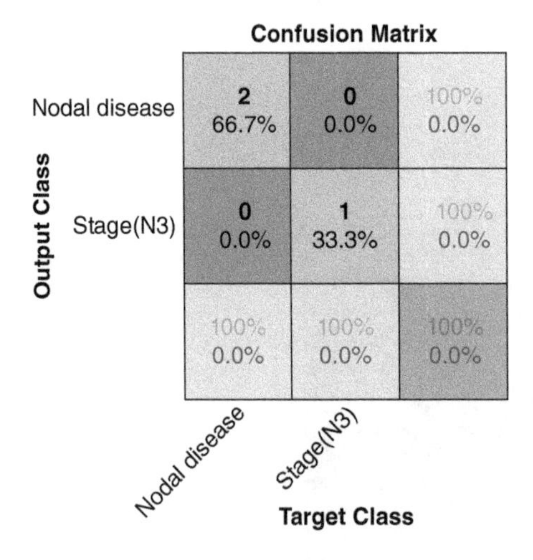

Figure 2.15 Confusion matrix for the lymph node (N).

2.4.4 Classification Accuracy of AlexNet

The validation frequency was set at three iterations, and the training progress for each class had a learning rate of 0.0001. The learning rate is kept lower for new classification tasks to achieve higher accuracy in less training time. The classification task was performed using a single CPU for computation. The experimentation results showed high classification performance with an accuracy of 94.44% for the primary tumor, followed by 100% each for the metastasis and lymph node, as shown in Table 2.2. The observation during the training progress was that an epoch of six produced a validation accuracy of 83.33% initially, and with each training process, there was an improvement in the accuracy with a maximum accuracy of 88.50%. A set with more epochs resulted in better performance of the network.

2.4.5 Comparative Analysis

Tables 2.2 and 2.3 provide insight into the performance evaluation of existing methods and the CNN architecture used in this research. TNM stands for the TNM staging procedure that includes Primary Tumor (T), Lymph Node (N), and Metastasis (M). The results depict the accuracy of the fine-tuned AlexNet with the use of the transfer learning approach. AlexNet has an edge over the performance of some of the studies listed in the table. The performance evaluation observation shows that most of the research was conducted using CT images, and the image processing task has segmentation and classification. Traditional approaches for image processing have been successful for detection but lack a scope for further improvement. The accuracy has been satisfactory for general segmentation and classification problems related to the detection of benign and malignant lung nodules.

Table 2.2 Performance evaluation of classification accuracy.

Lung Cancer Stages (TNM)	Validation Frequency	Hardware	Accuracy (%)
Primary Tumor	3 iterations	Single CPU	94.44
Metastasis	3 iterations	Single CPU	100
Lymph Node	3 iterations	Single CPU	100

Table 2.3 Comparative analysis of existing methods and the proposed method.

Authors	Data Set (Images)	Accuracy (%)
QingZeng Song et.al (2018)	CT	84.15
Siddharth Bhatia et al. (2018)	CT	84
Jiachen Wang et al. (2019)	CT	AUC:78.7
Yiwen Xu et al. (2019)	CT	DCNN:AUC:74,
Thomas Weikert et al. (2019)	PET-CT	T1, 2, 3, 4:90.4, 70.8, 29.4, 8.8
Margarita Kirienko et al. (2018)	FDG PET-CT	86.8, 69.3, 69.1
Ruo Xi Qin et al. (2020)	PET-CT	0.90, 0.5, 0.72
Ma et al. (2016)	CT	82
Proposed Architecture	PET-CT fusion	T:94.44, M:100, N:100

Table 2.4 Comparative analysis of the existing methods and the proposed method.

Authors	Methods
QingZeng Song et al. (2017)	DNN, CNN, SAE
Siddharth Bhatia et al. (2018)	Deep Learning, XGBoost
Jiachen Wang et al. (2019)	DCNN, XGB
Yiwen Xu et al. (2019)	CNN, RNN
Thomas Weikert et al. (2019)	Radiomics and Segmentation
Margarita Kirienko et al. (2018)	CNN
Ruo Xi Qin et al. (2020)	Deep Learning
Ma et al. (2016)	Random Forest
Proposed Architecture	Deep Learning

Furthermore, the studies that targeted TNM staging were mainly confined to the primary stage (T) with all the T parameters combined. The methods had higher accuracy in terms of the T1 stage, but the performance significantly dropped for subsequent stages to below a satisfactory clinical implementation level. The proposed architecture achieved higher accuracy concerning the TNM staging categories, which were comprised of all stages of the primary tumor (T), metastasis (M), and the lymph node (N). Such a network can be trained further with complex features for better results in the staging of cancer.

2.4.6 Limitations

There is a scope for improvement as deep learning technologies are rapidly becoming more advanced. Although AlexNet is a capable deep learning network, there have been further upgrades of the existing models in recent years that may produce superior results with more complex data. Therefore, a comparative analysis is required to understand the model's usability over more complex problems related to lung cancer classification. The proposed research has been conducted using a smaller data set. Therefore, the model has to be evaluated with a more complex data set to target advanced problems. Additionally, the feasibility of a deep learning model in a real-time clinical setting has to be evaluated as the challenges of high-computation power requirements may not be feasible in a medical facility.

2.5 Conclusion

The recent advancements in deep learning technologies are another gateway for delivering efficiency in medical imaging. Lung tumor detection is complicated due to the complexity of the organ and the clinical complications of lung cancer patients. Thus, accurate detection and classification is a difficult task. The complex lung tumor features, subcategories of diseases, and varying tumor sizes that refer to cancer stages are critical when focusing on the staging process. The classification of such lung cancer criteria might be limited for a traditional image processing task. CNNs have powerful computing capabilities to ensure that millions of images can be processed in a single iteration for a specific medical image processing problem. AlexNet is an efficient network for classification problems. However, its capabilities can be further evaluated with a larger data set with complex features. For the proposed research, a pretrained network was implemented with a transfer learning approach to achieve faster learning with significantly lower computational power. However, future research might extend to using other networks such as GoogleNet, DenseNet, VGG architecture, and AlexNet to compare the results focused on the staging process with a larger and more complex data set. Medical image processing will benefit immensely from using deep learning techniques that are capable of producing efficient results to aid doctors in achieving a successful clinical decision and treatment as well as eliminate misdiagnosis and treatment failures.

References

P. Angelus and J. Kirby. A large-scale ct and pet/ct dataset for lung cancer diagnosis (lung-pet-ct-dx). *Cancer Imaging Archive*, 2020.

S. Bhatia and S. Sinha. Lung cancer detection: A deep learning approach. *Soft Computing for Problem Solving*, 2018.

J. Brierley, M. Gospodarowicz, and B. O'Sullivan. The principles of cancer staging. *eCancerMedicalScience*, 2016. doi:10.3332/ecancer.2016.ed61.

J. Deng, W. Zeng, Y. Shi, and et al. Fusion of fdg-pet image and clinical features for prediction of lung metastasis in soft tissue sarcomas. *Computational and Mathematical Methods in Medicine*, 2020. doi:10.1155/2020/8153295.

A. Esteva, B. Kuprel, R. Novoa, and et al. Dermatologist-level classification of skin cancer with deep neural networks. *Nature*, 2017. doi:10.1038/nature21056.

S.H. Feng and S.T. Yang. The new 8th tnm staging system of lung cancer and its potential imaging interpretation pitfalls and limitations with ct image demonstrations. *Diagnostic and Interventional Radiology*, 2019. doi:10.5152/dir.2019.18458.

P. Goldstraw, K. Chansky, J. Crowley, and et al. The iaslc lung cancer staging project: Proposals for revision of the tnm stage groupings in the forthcoming (eighth) edition of the tnm classification for lung cance. *Journal of Thoracic Oncology*, 2016. doi:10.1016/j.jtho.2015.09.009.

B. Hochhegger, G. Alves, K. Irion, and et al. Pet/ct imaging in lung cancer: indications and findings. *Jornal Brasileiro de Pneumologia*, 2015. doi:10.1590/S1806-37132015000004479.

A. Hosny, C. Parmer, J. Quackenbush, and et al. Artificial intelligence in radiology. *Nature Reviews Cancer*, 2018. doi:10.1038/s41568-018-0016-5.

S. Huang, Y. Jie, S. Fong, and Q. Zhao. Artificial intelligence in cancer diagnosis and prognosis: Opportunities and challenges. *Cancer Letters*, 2019. doi:10.1016/j.canlet.2019.12.007.

M. Jingchen, Q. Wang, Y. Ren, H. Hu, and J. Zhao. Automatic lung nodule classification with radiomics approach, year: 2016, journal: Medical Imaging 2016, PACS and Imaging Informatics: Next Generation and Innovations, doi:10.1117/12.2220768.

M. Kirienko, M. Sollini, G. Silvestri, and et al. Convolutional neural networks promising in lung cancer t-parameter assessment on baseline fdg-pet/ct. *Contrast Media & Molecular Imaging*, 2018. doi:10.1155/2018/1382309.

R. Kochhar. Pet-ct in staging lung carcinoma. *The Christie NHS Foundation Trust*, 2019.

K. Kourou, T. ExarcosP, K. ExarcosP, and et al. Machine learning applications in cancer prognosis and prediction. *Computational and Structural Biotechnology Journal*, 2015. doi:10.1016/j.csbj.2014.11.005.

A. Krizhevsky, I. Sutskeve, and G. Hinton. Imagenet classification with deep convolutional neural networks. neural information processing systems. *Neural Information Processing Systems*, 2012. doi:10.1145/3065386.

G. Litgens, T.K.B.E. Bejnordi, A. Setio, and et al. A survey on deep learning in medical image analysis. *Medical Image Analysis*, 2017. doi:10.1016/j.media.2017.07.005.

R. Paul, S. Hawkins, Y. Balagurunatham, and et al. Deep feature transfer learning in combination with traditional features predicts survival among patients with lung adenocarcinoma. *MDPI: Tomography*, 2016. doi:10.18383/j.tom.2016.00211.

Y. Pinilla, B.R. Vigil, and N.G. Leon. Integrated fdg pet-ct: Utility and applications in clinical oncology". *Clinical Medicine Insights: Oncology*, 2008. doi:10.4137/cmo.s504.

V. Purandare, and N.C. Rangarajan. Imaging of lung cancer: Implications on staging and management. *Indian Journal of Radiology and Imaging*, 2015. doi:10.4103/0971-3026.155831.

R.X. Qin, Z. Wang, L. Jiang, and et al. Fine-grained lung cancer classification from pet and ct images based on multidimensional attention mechanism. *Complexity*, 2020. doi:10.1155/2020/6153657.

M.W. Saif, I. Tzannou, N. Makrilia, and K. Syrigos. Role and cost effectiveness of pet/ct in management of patients with cancer. *Yale Journal of Biology and Medicine*, 2010.

T.K. Sajja, R.M. Devarapalli, and H.K. Kalluri. Lung cancer detection based on ct scan images by using deep transfer learning. *IIETA*, 2019. doi:10.18280/ts.360406.

Q.Z. Song, L. Zhao, X.K. Luo, and et al. Using deep learning for classification of lung nodules on computed tomography images. *Journal of Healthcare Engineering*, 2017. doi:10.1155/2017/8314740.

A. Teramoto, A. Yamada, Y. Kiriyama, and et al. Automated classification of benign and malignant cells from lung cytological images using deep convolutional neural network. *Informatics In Medicine Unlocked*, 2019. doi:10.1016/j.imu.2019.100205.

J. Wang, R. Gao, Y. Huo, and et al. Lung cancer detection using co-learning from chest ct images and clinical demographics. *Proceedings SPIE 10949, Medical Imaging 2019: Image Processing, 109491G*, 2019. doi:10.1117/12.2512965.

T. Weikert, T.A. D'Antonoli, J. Bremerich, and et al. Evaluation of an ai-powered lung nodule algorithm for detection and 3d segmentation of primary lung tumors. *Contrast Media & Molecular Imaging*, 2018. doi:10.1155/2018/1382309.

B. Wendy, A. Hosny, M. Schabath, and et al. Artificial intelligence in cancer imaging: Clinical challenges and applications. *CA: A Cancer Journal for Clinicians*, 2019. doi:10.3322/caac.21552.

D. Wu, C. Yang, S. Wong, and et al. An examination of tnm staging of melanoma by a machine learning algorithm. *2012 International Conference on Computerized Healthcare (ICCH), IEEE*, 2012. doi:10.1109/ICCH.2012.6724482.

X.U. Yiwen, A. Hosny, R. Zelenik, and et al. Deep learning predicts lung cancer treatment response from serial medical imaging. *Cancer Research Communications*, 2019. doi:10.1158/1078-0432.

3

Formal Methods for the Security of Medical Devices[1]

Srinivas Pinisetty[1], Nathan Allen[2], Hammond Pearce[3], Mark Trew[4], Manoj Singh Gaur[5], and Partha Roop[6]

[1] *Indian Institute of Technology Bhubaneswar, India*
[2] *University of Auckland, New Zealand*
[3] *NYU Tandon School of Engineering, USA*
[4] *University of Auckland, New Zealand*
[5] *Indian Institute of Technology Jammu, India*
[6] *University of Auckland, New Zealand*

3.1 Introduction

Cyber-physical systems (CPSs) usually consist of a network of embedded devices that control adjoining physical processes (Alur, 2015). Numerous examples, both visible and invisible, surround our modern lives: smart grids manage and monitor national power grids, intelligent transportation systems negotiate city traffic flows, industrial robotic systems produce high volumes of precision products, and implantable medical devices (IMDs) to overcome disease and disabilities. Since many of these applications serve to enhance our over overall quality of life there is the potential for catastrophic consequences when such systems fail. These kinds of CPSs are *safety-critical*: for us to be safe, they must operate correctly.

There is a large body of work on examining and ensuring the safety of CPSs (Alur, 2015). Breaches can occur due to the random failures of integrated sensors, actuators, or other hardware and software components that control the system. To mitigate the consequences of system failures, functional safety standards such as IEC 61508 (Bhatti et al., 2016) or its variants can be used. These provide for certification of the electrical, electronics, and programmable-electronic components of a CPS. They demand systematic risk analysis to consider the failure of hardware components along with the development of associated mitigation strategies. Software components, on the other hand, which can suffer from other kinds of failures, require alternate testing, such as the "Vee" model (Forsberg and Mooz, 1991).

*Corresponding Author: spinisetty@iitbbs.ac.in
[1]This work has been partially supported by The Ministry of Human Resource Development, Government of India (SPARC P#701), IIT Bhubaneswar Seed Grant (SP093).

Applied Smart Health Care Informatics: A Computational Intelligence Perspective, First Edition.
Edited by Sourav De, Rik Das, Siddhartha Bhattacharyya, and Ujjwal Maulik.

While the certification of CPSs is critical for ensuring system safety, failures leading to safety violations that arise from security breaches are not in the purview of these safety standards. Hence, many certified CPSs have experienced serious failures due to cyber-physical attacks (CP attacks) launched with malicious intent (Loukas, 2015). This class of attacks, which originate in cyber space, have a very real potential for causing major, physical real-world damage via the deliberate violation of safety guarantees. Some notable examples over the past two decades include the Maroochy Shire wastewater attack (Weiss, 2014), the Los Angeles traffic system hack (Kelarestaghi et al., 2018), the Turkish pipeline explosion (Madnick and Mangelsdorf, 2017), the Stuxnet worm (Chen, 2010), and the Tesla remote hack (Nie et al., 2017).

Given the potentially catastrophic impact of CP attacks, there have been many recent attempts at attack detection and mitigation. This is complicated by the requirements of devices in the CPS domain, which are often not suited to the conventional methods of cyber-security (typically founded on number-theoretic encryption). Hence, one recent work focused on light-weight encryption over the IEC 61499 standard for Industrial automation (Tanveer et al., 2020). Other approaches focus on improving the devices to provide more guarantees. For instance, "smart" input/output (I/O) modules for industrial programmable logic controllers (PLCs) have an awareness of the safety requirements their application is developed in to mitigate CP attacks in industrial control systems (Pearce et al., 2019a). Moreover, there have been many works examining CP attacks on smart grids, and a recent survey provides a good summary (He and Yan, 2016).

Formal methods (Woodcock et al., 2009; Baier and Katoen, 2008; Alur, 2015) constitute a class of mathematically-founded techniques for the rigorous verification of safety critical systems. Consequently, they are ideal when dealing with attack detection and mitigation of CP attacks. Application examples of formal techniques include Chen et al. (2017), Pasqualetti et al. (2013), which use control theory, and those which are based on formal methods (Lanotte et al., 2018, 2017). In this chapter, we will detail the formal methods applied to the detection and mitigation of CP attacks on medical devices.

Here, we consider medical devices such as implantable cardioverter defibrillators (ICDs), cardiac pacemakers, insulin pumps, and gastric electrical stimulators (GESs). These devices are used to control a specific disease by continuously sensing signals from a given organ. In the event of abnormal activity, the device provides appropriate correcting input to the organ through an actuator. Hence, these devices act as controllers, while the organ in question acts as its adjoining plant (i.e. the physical process being controlled). For example, consider a pacemaker or ICD: the plant is the patient's heart, while the controller is the device itself. The interface is through electrical leads, which are connected inside the chambers of the heart and act as both sensors and actuators.

As discussed in the survey (Camara et al., 2015), implantable medical devices have been extensively studied to ensure their safety. These medical devices are usually partially or completely contained within the human body, and hence any failure may be exceedingly complicated to resolve, including the need for surgery, which introduces further risk for the life of the patient. Modern enhancements, such as wireless-enabled operation can facilitate many benefits, including device reprogramming and customization using wireless programmers. Bed-side units also use wireless communication to manage patient data for remote analysis by the physician. Patients and healthcare workers may have the functionality to access the device remotely using a mobile phone application. However, such wireless operations also expose these devices to different types of CP attacks (P. and Aziz, 2018): pacemakers that give deadly shocks (Kirk, 2012), insulin pumps programmed to deliver unsafe levels of insulin (Thompson et al., 2018), and denial of service (DoS) attacks on ICD devices (Marin et al., 2016). Such attacks are also frequently reported in the media, for example, the Wall Street Journal cited Food and Drug Administration (FDA) report that said "we are aware of hundreds of medical devices that have been infected by malware" (Weaver, 2013).

Considering the immense challenge of CP attacks on medical devices, we consider the cardiac pacemaker case study as a motivating example in the next section to illustrate the problem of CP attacks. While our focus is on pacemakers, the discussions and specific methods we present are generic to most medical devices that act as controllers for specific organs.

3.1.1 Pacemaker Security

Cardiac pacemakers are used to treat arrhythmia, a condition where the patient's heart produces irregular heartbeats. During an arrhythmia, the heart can beat too fast, which is called *tachycardia*, or too slow, which is called *bradycardia*. The pacemaker is implanted under the skin of the patient's chest, typically just under the collarbone, and connected to the heart using conductive and biologically inert leads. The pacemaker senses intrinsic *atrial* and *ventricular* events in the heart and gives electrical pacing pulses (either an atrial pulse or a ventricular pulse) as necessary, which controls the heart in a closed-loop.

Like many other medical devices, wirelessly-accessible pacemakers are common, and many CP attacks have been reported (Sametinger et al., 2015). Hackers can change the rate of the pacemaker, causing insufficient or overly rapid pacing. Rapid pacing is particularly dangerous and can initiate tachycardia or fibrillation, which have a high likelihood of sudden cardiac death. Other attacks have demonstrated that pacemaker batteries can be drained at an alarming rate. In the detailed Sametinger et al. (2015) survey, several possible hacking avenues for pacemakers were reported, and specific instances were reported in Kirk (2012) and Clery (2015).

A key problem for pacemaker security is the trade-off of allowing device access to health care professionals for emergency situations while ensuring that the device prevents unauthorized access. This is discussed in more detail in the survey by AlTawy and Youssef (2016). While cryptography is one of the best mechanisms for securing such CPS, a significant challenge is key distribution to legitimate parties, such as emergency health care workers, while preventing nefarious third parties from accessing those same keys.

Considering this, alternative solutions based on devices that automatically detect and respond to anomalous behavior have been developed (Zhang et al., 2013). MedMon (Zhang et al., 2013) can detect and prevent anomalous events by snooping all wireless transactions. MedMon relies on a set of policies to monitor and determine which events need to be jammed. MedMon monitors a set of policies to identify events for jamming. However, this approach has limitations. Additional embedded devices with associated costs, power consumption, and certification needs are required to snoop the wireless channels. The effectiveness depends on the defined policies, and as these are informal, there is no guarantee of soundness in the developed framework. In comparison, we present an approach that has soundness guarantees provided by techniques grounded in formal methods (Woodcock et al., 2009).

3.1.2 Overview

The remainder of this discussion is organized as follows. In 3.2 we introduce the pacemaker case study in detail, including its closed-loop operation and associated timing requirements. In 3.3 we recap the formal methods relevant for this discussion. Then, in 3.4 we present two formal techniques for attack detection and mitigation for use with pacemakers. Finally, in 3.5 we make concluding remarks and motivate the need for further research on several topics not yet addressed to prevent and mitigate the effects of CP attacks.

3.2 Background: Cardiac Pacemakers

Pacemakers are used to both sense the activity of the cardiac system (the heart) and provide pacing signals when necessary – for instance when bradycardia occurs and the heart is beating too slowly. These devices are implanted under the patient's skin and interact through leads that connect the pacemaker to the internal walls of the heart, as shown in Figure 3.1. In a more complex scenario, leads are placed in both ventricular chambers and bi-ventricular pacing is used to synchronize beat timing across the heart.

The interaction of a digital pacemaker and physical heart is an instance of a safety-critical CPS – as any fault could have catastrophic consequences, including death. To prevent this, pacemakers must operate correctly according to a set of strict timing properties and intervals. Therefore, these systems may also be characterized as *hard real-time systems*. In other words, both the functionality of the device (that it can pace) is just as important as the timing correctness of the device (when it should pace). As will be discussed later, this is applicable to runtime enforcement (RE) and Runtime verification (RV).

Society is becoming increasingly reliant on cardiac pacemakers for their health benefits. A 2009 survey found that 737 840 new implantations and 264 824 replacements were performed in that year, for a total of 1 002 664 implantations (Mond and Proclemer, 2011). However, these devices are not without their issues. There were

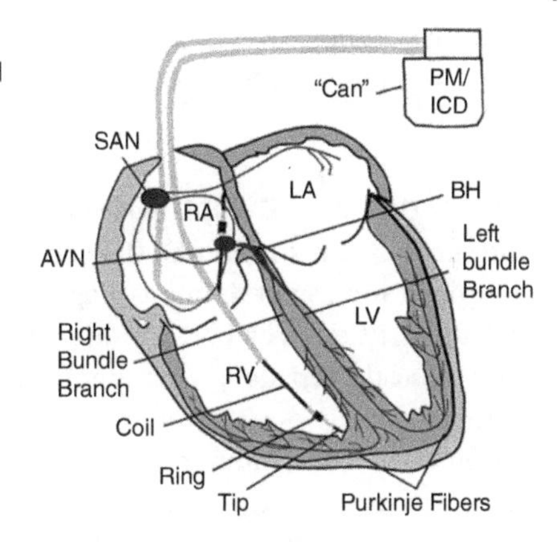

Figure 3.1 The heart-pacemaker system shows leads connected to the right atrium and ventricle (reproduced from Pinisetty et al. (2018)). (PM/ICD - pacemaker/implantable cardio-defibrillator; SAN - sino-atrial node; LA - left atrium; RA - right atrium; AVN - atrio-ventricular node; LV - left ventricle; RV - right-ventricle).

over 210 000 recorded ICDs and pacemakers recalled between 2006 and 2011 due to various issues – some with immediate safety concerns, while others were more precautionary (Alemzadeh et al., 2013).

Due to their implanted nature, the impact of these recalls is not just isolated to the device, but also requires costly (and potentially dangerous) operations to remove and/or replace the device. The additional costs due to replacements and the recall for the above five-year period is close to one billion dollars (Maisel et al., 2001). Further, the malfunctions in the group of recalled devices for 2006 to 2011 led (directly and indirectly) to over 17 500 injuries and 350 deaths (Alemzadeh et al., 2013).

3.2.1 Pacemakers

Fundamentally, a pacemaker is a controlled pulse generator that is able to observe the natural behavior of the heart in order to either enable or disable its own output. The simplest form of pacemaker is one that has a single lead (over which both sensing and pacing are performed) that is usually placed in the right ventricle. However, dual-chamber pacemakers (with an additional lead in the right atrium) are much more commonplace today, and are able to correct more adverse behaviors such as atrium-ventricle synchronization.

The different variations of pacemaker are generally identified through a three or four character code. The first and second characters specify where the pacing and sensing are performed by the pacemaker, and can either be "A" (atrium), "V" (ventricle), or "D" (dual, or both). The third character refers to the logic that the device uses to decide when pacing signals should be sent. A value of "I" (inhibited) means that the pacing event is stopped by the presence of a natural event, which otherwise would be sent. Alternatively, "T" (triggered) means that the pacing event will be generated after some delay of a natural event (e.g. to maintain synchronization). There is also a "D" (dual) mode, which includes logic of both the inhibited and triggered modes.

In this work, we will be looking at a DDD mode pacemaker, meaning that it senses and paces in both the atrium and the ventricles (i.e. it is a dual lead/dual chamber pacemaker) and supports both inhibited and triggered operations. Other common forms of pacemakers typically only use a single lead and are generally either VVI for pacing the ventricles or AAI for pacing the atria, depending on the patient's condition.

3.2.1.1 Operation of a DDD Mode Pacemaker

The simplest way to describe the operational behavior of a DDD mode pacemaker is through the use of a timing diagram. This diagram is shown in Figure 3.2 and includes the signals on both the atrial and ventricular leads, known as electrograms (EGMs), at the top along with the timers used to create the pacemaker logic at the bottom.

Figure 3.2 shows four beats of the heart, labelled ❶ through ❹. The EGMs traces at the top of the figure have three different types of events that consist of natural events (atrial sense (AS) and ventricle sense (VS)), refractory (ignored) events (atrial refractory sense (AR) and ventricle refractory sense (VR)), and pacing events (atrial pulse (AP) and ventricle pulse (VP)). It is also important to note that the sign of the pacing events (AP and VP) are opposite to the sensed events due to the mechanism that pacemakers use to stimulate the activity.

The bars in the lower portion of Figure 3.2 represent the operation of six timers that are used to create the pacemaker's internal logic. ventricular refractory period (VRP) and post-ventricular atrial refractory period (PVARP) are two timers used to ignore either natural ventricular or atrial events, respectively. The effect of these timers is seen in the figure, where an AR is created during the PVARP of ❶ and a VR during the VRP of ❷.

To ensure synchronization between the atria and ventricles, there exist two additional timers – atrioventricular interval (AVI) and atrial escape interval (AEI). AVI ensures the correct synchronization between atrial events and their subsequent ventricular events. This can be seen by the AVI timer that starts on all atrial events, and if the timer should ever expire without a VS, such as beat ❷, then a VP is generated. AEI, on the other hand, is the reciprocal of this whereby it is started on

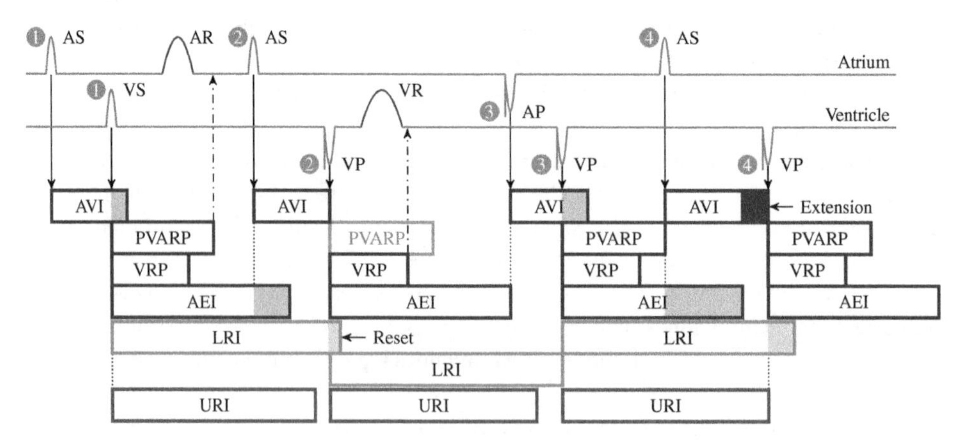

Figure 3.2 Timing diagram for a DDD mode pacemaker (adapted from Pinisetty et al. (2017c)).

every ventricular event, and if no AS occurs, it generates an AP on expiration, such as in beat 2.

Finally, the two bottom-most timers of Figure 3.2 are lower rate interval (LRI) and upper rate interval (URI), which govern the limits on the overall allowed pacing rates for the pacemaker. LRI is the lowest rate (i.e. highest interval) that is allowed for ventricular events. When this timer expires (as in beat 3) a VP will be generated. URI is the highest rate (i.e. shortest interval) that is ever allowed for ventricular pacing events from the pacemaker. While this timer is running, no VPs are generated from the device, and instead, these events are delayed until the expiration of the URI timer. This effect can be seen in beat 4 where the AVI timer is extended until URI has expired before emitting the VP event.

3.2.2 The Cardiac System

The cardiac system of the human body is centered around the heart, the electrically-driven biological pump that facilitates the movement of deoxygenated blood to the lungs and oxygenated blood throughout the rest of the body. The heart achieves this through a series of four chambers – two atria and two ventricles, with one of each being on the "right" of the heart and the other on the "left". The atria are the input chambers where blood is received either from the body (right atrium) or the lungs (left atrium). This blood is then pumped from the atria down to their respective ventricles through contraction of the cardiac muscle.

Once filled, the ventricles then perform their own, much larger contractions to force the blood out of the heart and to either the lungs (right ventricle) or the rest of the body (left ventricle). The delay between atrial and ventricular contractions is critical to ensure optimal cardiac output. If the delay is too short, the ventricles will not be completely filled from the atria, which reduces the blood pumped per beat. Alternatively, if the delay is too long, the cardiac output becomes more periodic, which reduces the flow rate.

At a high level, there are two main electrical control regions of the heart that we will describe here. Firstly, the sinoatrial (SA) node, located near the middle of the atria, acts as the *natural pacemaker* of the heart. This region electrically cycles autonomously at a rate of 60 to 100 beats per minute (bpm), which initiates a single cycle of the heart each time. The electrical stimulation propagates to nearby cells, firstly through each of the atria before travelling to the ventricles. Should this autonomous behavior of the SA node ever fail, then the region of the heart with the next highest intrinsic rate will take over these pacemaking duties.

The path between the atria and ventricles is controlled by the atrioventricular (AV) node, which serves two purposes. First, this node is the main region that creates delay to ensure optimal synchronization between atrial and ventricular contractions. Therefore, if this region of the heart has a conduction block that stops propagation from occurring, then synchronization between the atria and ventricles is lost, which requires some external (i.e. pacemaker) action. Additionally, the AV node serves as the secondary autonomous region of the heart, is capable of pulsing at a rate of 40 to 60 bpm, and will take over in the case of SA node failure. However, the location of this node, being more central to the heart, results in a vastly reduced delay between atrial and ventricular contractions. The slower intrinsic rate also results in a slower heart rate.

Should the autonomous behaviors of the AV node fail, then the region with the next highest intrinsic rate are located even further down the conduction network, close to the ventricles that pulse at a rate between 20 and 40 bpm. The location of this next region produces *inverse propagation* through the heart where synchronization between the atria and ventricles is essentially lost; however, it is still able to maintain some minimal cardiac function and only serves as backup in a worst-case scenario.

3.2.2.1 Electrograms and Electrocardiograms

The activity of the heart is generally captured through two different signals that depend on the reading location. These signals use electrodes to measure the electrical potential of the heart as cells go through their contraction (depolarization) and relaxation (re-polarization). As cells along the conduction path, from atria to ventricles, go through this cycle the electrodes detect the change in potential for various regions of the heart.

For devices that have leads embedded on the inside surface of the heart, such as pacemakers, these signals are known as electrograms (EGMs). In this case, the proximity of the electrode to the cardiac cells results in a signal that is very strongly weighted to the area around where the electrode is placed. Through the use of multiple leads, such as for a DDD mode pacemaker, the activity of multiple regions can be detected.

Alternatively, most noninvasive measurements of the heart use electrodes placed on the body surface at various points to record what is known as electrocardiograms (ECGs). The most common example of this is the standard 12-lead ECG, which uses 10 electrodes placed on the limbs of the patient and the chest around the heart. The 12 leads are comprised of nine *physical leads*, which are simply the potential of an electrode relative to the ground electrode, and three *virtual leads*, which are created using the difference between various limb electrodes. Despite being further away from the cardiac cells, these 12 leads of an ECG are able to capture the heart activity in relatively high detail, which allows physicians to diagnose cardiac diseases in a noninvasive manner.

A typical ECG signal is illustrated in Figure 3.3, with annotations that indicate the various regions of the signal. Firstly, the P wave (Ⓟ) represents the atrial depolarization, which indicates that the atria have started to contract. Next, the QRS complex, which generally consists of a positive deflection (Ⓡ) surrounded by negative deflections (Ⓠ and Ⓢ), captures the ventricular depolarization, and immediately precedes when the ventricles begin to contract to pump blood around the body. Finally, the T wave (Ⓣ) is caused by ventricular re-polarization, where the ventricles have once again relaxed.

From these points there are a few, key timing intervals with measurable physical significance. The PR interval is measured as the time between the start of the P wave and the start of the subsequent QRS complex. This is a measure of the delay between atrial and ventricular depolarizations (i.e. the AV interval). Additionally,

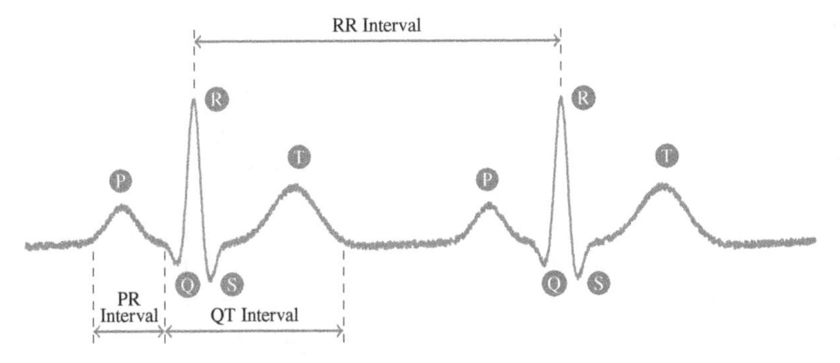

Figure 3.3 Timing information of electrocardiogram signals.

the RR interval is, as the name implies, simply the difference between successive R waves. This interval corresponds to the time between successive heart beats, which is the inverse of the heart rate.

Finally, the last key interval of the an ECG is the QT interval, which is measured as the time between the start of the QRS complex and the end of the subsequent T wave. The QT interval captures the time from ventricular depolarization to re-polarization, which is a measure of the action potential duration (APD) of the ventricular cells. Due to the restitution behavior of the underlying cardiac cells, this interval can change depending on the heart rate. To correct this, a *corrected QT interval* (QT_c) is often used and is calculated using the RR interval per Equation 3.1.

$$QT_c = \frac{QT}{\sqrt{RR}} \tag{3.1}$$

3.3 State of the Art, Formal Verification Techniques

In 3.2, we provided an overview of the cardiac pacemaker and the associated timers of a DDD-mode pacemaker. As is evident, the pacemaker works in close-loop with the heart to enforce the correct timing of cardiac electrical activity. Consequently, this is a safety-critical CPS. This section provides an introduction to formal methods (Woodcock et al., 2009), a collection of techniques which are ideal for ensuring the safety of CPS. We first provide a short overview of various techniques, related to checking the correctness of a system (that is checking whether the system satisfies the requirements). Later, we describe the runtime enforcement (RE) technique, which is related to correcting the execution of a system (at runtime) according to the requirements.

Generally, requirements are informal descriptions of what we want a system to do. Requirements are the basis for the design, development, and testing of a system. When requirements are captured in some representation, using either natural language or a (semi-)formal representation such as unified modelling language (UML) (Chaudron et al., 2012), it is known as the *system specification*. The process of checking the correctness of a system (that is, whether the system confirms to the requirements) is called *verification*. A form of verification is achieved through testing. For testing, the specification of the system acts as the correctness criterion against which the system is compared. A common practice is to write the specification informally for this purpose, generally in natural language.

Testing is the process of executing a system with the objective of detecting errors. Errors happen when the system's output trace deviates from the desired output as captured in the specification. Generally, testing is performed by executing the system in a controlled environment, providing test data as an input to the system, and making observations about its outputs. From the observations made during the execution, verdicts about the correctness of the system (whether it satisfies the specification) can be made.

Drawbacks of Testing-Based Verification Although informal specifications written in a natural language such as English are easy to read, they may be ambiguous. Informal description of requirements are often incomplete and liable to different, and possibly inconsistent, interpretations. Relying on unclear, imprecise, incomplete, or ambiguous specifications causes several problems in the testing process, which often makes these informal specifications inappropriate. While specifications in semi-formal standards such as UML are better, they are still not entirely formal and are usually outside the purview of formal analysis. Importantly, testing can be used to find the presence of errors but can not be used to prove their absence, which is critical to ensuring the safety of a CPS.

Formal Specification and Formal Methods A specification is said to be formal if it is expressed with a precise syntax and semantics as well as every statement having a unique meaning, mathematically. These are called formal languages. Such formal specification languages are often based on concepts such as mathematical logic. The advantage of such formal specifications and models is that they have precise, unambiguous semantics, which enables an analysis of systems and reasoning with mathematical precision.

Thus, using formal languages to specify requirements removes ambiguities and inconsistencies, which allows them to be used as contracts between developers and customers. Moreover, formal languages are more easily amenable to tool-based automatic processing. In contrast to informal specifications, formal specifications can be mathematically analysed since they are mathematically based. For example, methods (and tools) exist to automatically verify the absence of deadlock in the formal description of the design of the system. Formal specifications can also be used to generate test cases automatically.

3.3.1 Formal Verification Techniques

Verification is the process of checking whether a system or product satisfies (conforms to) its specification. When formal languages and techniques (based on mathematics) are used in the verification process, this is known as *formal verification* (Baier and Katoen, 2008). Formal verification techniques can be divided into two subcategories called static and dynamic verification techniques.

3.3.1.1 Static Verification Techniques

Static verification techniques are performed without actually executing the system that is being verified. Some of the static formal verification techniques include model checking, static analysis, and theorem proving.

Model Checking Model checking (Baier and Katoen, 2008) is an automated approach to verify that a formal model of the system, usually described as a state transition system, satisfies the formal specification that defines the requirements of the system. Generally, the model of the system will be an abstraction, which omits details irrelevant for checking the desired properties and describes how the state of the system may evolve over time. These are typically described as automata (or their extensions with time, probability, cost, data, etc.), which capture the possible states, initial state, and the possible transitions between states. Properties are usually expressed in some form of temporal logic such as linear temporal logic (LTL) or computational tree logic (CTL) (Huth and Ryan, 2004), which have constructs to express constraints on how the state of a system may evolve. Model checkers such as Spin (Holzmann, 1997) and UPPAAL (Larsen et al., 1997) are examples of popular tools for model checking.

Figure 3.4 shows the process of applying model checking. A model of the system is created and the requirements, which describe the desired properties of the system informally, are formalized. These formalized requirements are called properties (or property specifications). The model-checking algorithms (and tools), take the model and properties as input, and determine whether the model satisfies the properties or not. The model checker explores all the possible (reachable) states to decide whether the properties hold. In the case where a property does not hold, the model checker returns a counter example, which is a run of the model of the system that leads to a state in which the property does not hold.

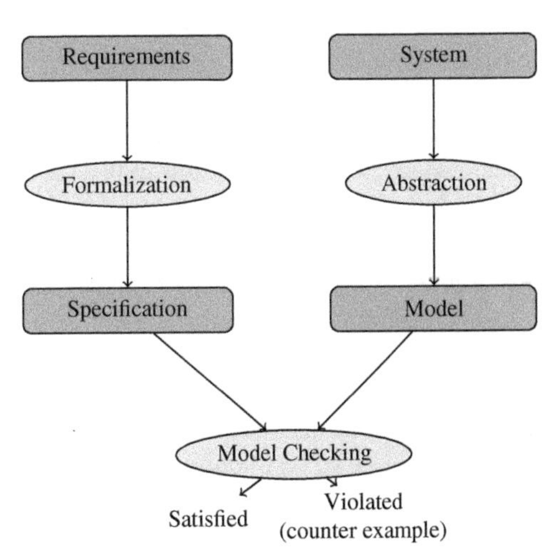

Figure 3.4 Model checking.

Static Analysis Static analysis is a powerful technique that enables automatic verification of programs. Using a static analysis approach, various properties such as type safety and resource consumption can be checked (Hubert et al., 2011). Abstract static analysis (D'Silva et al., 2008) techniques, which are mostly used for compiler optimization, are also used for program verification. These techniques focus on computing approximate *sound* guarantees efficiently. However, the information provided is not always precise since it is based on an approximation. Generally, static analysis tools such as Evans and Larochelle (2002), Hubert et al. (2011) parse the source code and build an abstraction (model) of the system such as a directed graph. By traversing the model, it checks whether certain properties hold in the model, and if a violation is found in the model, it can also be expected in the source code.

Theorem Proving In theorem-proving approaches, for a mathematical statement to be true, a convening mathematical proof is constructed. If a proof cannot be found for a mathematical statement, then it cannot be concluded that the statement is true. In the case that a correct proof is found, the statement is called a theorem. Several tools exist to help in the process of constructing formal proofs. For example interactive theorem provers such as COQ (Bertot and Castéran, 2020) can be used to construct a proof interactively. Proof checkers such as Met can be used to check whether a proof is correct or not.

3.3.1.2 Dynamic Verification Techniques

Verification techniques such as model checking, static analysis, and theorem proving verify the overall system exhaustively. When a proof is done, then it guarantees 100 % coverage for a specific property. However, there are issues such as scalability that affect these static-verification methods. Hence, light-weight verification techniques, which perform verification through system execution, are considered complimentary approaches. Dynamic verification techniques execute the system in a particular environment, provide specific input data to the system, and observe its behavior (or output).

Testing Using Formal Methods Several problems occur in the testing process since the specifications, when written informally, are often incomplete or ambiguous. Basing testing activities on a formal specification is an advantage since formal specification is precise, consistent, and unambiguous. This is the first big advantage in contrast to traditional testing processes where such a basis for testing is often lacking.

Checking the functional correctness of a black-box system by means of testing is known as conformance testing (Jan, 2002; Jard and Jéron, 2005). Formal specifications are also amenable to automatic processing by tools, which is another advantage of their use for testing. For conformance testing with formal methods, a formal specification is the starting point for the generation of test cases (Jan, 2002). Figure 3.5 gives an overview of the process. Algorithms (which have sound formal basis) have been developed to automatically generate test cases from a formal specification. This opens the way towards completely automated testing,

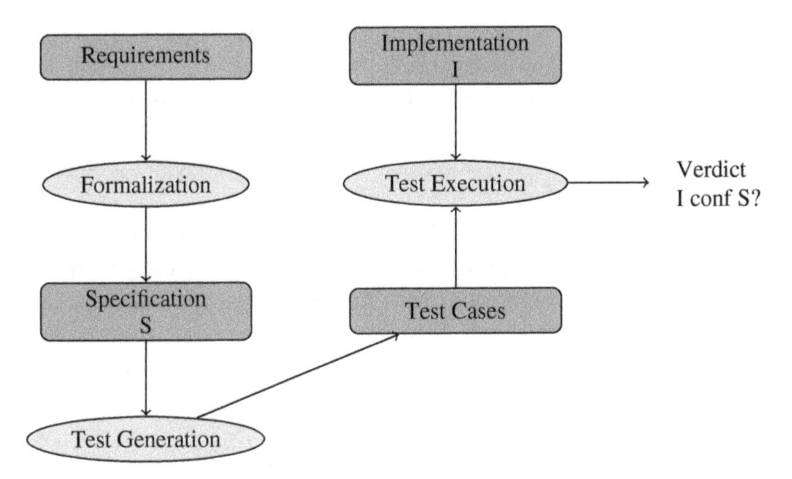

Figure 3.5 Conformance testing with formal methods.

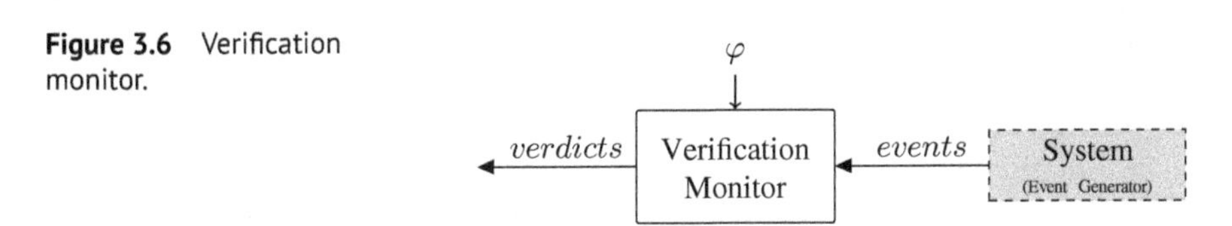

Figure 3.6 Verification monitor.

where the system under test and its formal specification are the only prerequisites. Moreover, they have been implemented in tools such as TGV (Jard and Jéron, 2005), STG (Clarke et al., 2002), and TorX (Bohnenkamp and Belinfante, 2005), which lead to the automatic, faster, and less error-prone generation of test cases. STG avoids enumerating the specification's state space and uses symbolic generation techniques. Regarding the conformance testing of real-time systems, there are tools such as TTG (Krichen and Tripakis, 2009), which is based on the model of partially-observable, non-deterministic timed automata.

3.3.2 Runtime Verification

Runtime verification (RV) (Falcone and Zuck, 2012; Bauer et al., 2011) approaches allow checking whether an execution of a system under observation satisfies (or violates) a desired correctness property. RV is a light-weight formal verification technique and it complements the other formal verification techniques discussed so far, such as model checking.

RV techniques do not influence the program execution, and deal only with the detection of violation (or satisfaction) of properties. In RV, a monitor is used to check whether the run of a system satisfies a property.

In RV approaches, the system being monitored is treated as a sequence generator. As illustrated in Figure 3.6, a verification monitor is a decision procedure that emits verdicts, which provide information regarding the correctness of the (partial) observed trace generated from the system execution.

Formally, if φ denotes the property (i.e. the set of valid runs), then the RV problem is to check whether a system run belongs to the property φ. The set of verdicts may contain more than two different truth values such as {*true*, *false*, *inconclusive*}. The verdict provided by the monitor is *true* if the current input fulfils the property (irrespective of how the current input is extended), *false* if a misbehavior is detected, and *inconclusive* otherwise.

The monitor can operate offline (by reading a log or sequence of system events and/or actions), or online (by receiving events in a lock-step manner with the execution of the system). Most of the formal RV approaches deal with the synthesis of RV monitors from some high-level specification of the property that the monitor should verify. Monitors are synthesized from specifications such as LTL (Bauer et al., 2011).

3.3.2.1 A Brief Overview of Some Runtime Verification Frameworks

RV frameworks can be distinguished based on the to the formalism used to express the desired properties. Monitoring propositional properties has received a lot of attention (Bauer et al., 2011). In propositional approaches, properties refer to events taken from a finite set of propositional names, such as one that defines the ordering of function calls in a program.

Parametric approaches such as Chen and Rosu (2009) and Barringer et al. (2012) have received growing interest in the past few years. In this case, events in the property are augmented with formal parameters, which are instantiated at runtime.

In timed approaches (Pinisetty et al., 2012), the time between events may influence the truth-value of the property. Monitoring real-time properties (i.e. where time is continuous) is a more complex and harder problem. Firstly, modelling timed specification requires a formalism that involves time as a continuous parameter, which adds complexity. Second, when monitoring a timed property, the time spent executing the monitoring code, i.e. the overhead induced by the monitor, influences the produced verdict of the monitored property.

The following are some of the approaches proposed for monitoring systems with respect to timed properties. In Bauer et al. (2011), an approach for RV of time-bounded properties expressed in a variant of timed linear temporal logic (TLTL) is proposed. The truth-values of the considered logic (called TLTL_3) belong to the set $\{\top, \bot, ?\}$, where the verdicts true ($\top$) and false ($\bot$) are conclusive, and (?) is an inconclusive verdict. After reading some timed word u, the monitor synthesized for a TLTL_3 formula φ states verdict $\top$ (or $\bot$), when there is no infinite timed continuation, w, such that $u \cdot w$ satisfies (or does not satisfy) φ. Metric temporal logic (MTL) (Nickovic and Piterman, 2010) is another variant of LTL in a timed context. Nickovic and Piterman (2010) proposed a translation of MTL to timed automata. The translation is defined under certain assumptions, e.g. a bounded number of events can arrive to the monitor in a finite interval. In Thati and Rosu (2005), an online monitoring algorithm is proposed that uses a rewriting approach of the monitored formula. The authors also study its complexity. Basin et al. (2011) proposed an improvement of the aforementioned approach with better complexity but considered only the past fragment of MTL. Other works related to

RV of timed properties are tools such as LARVA (Colombo et al., 2009) and AMT (Nickovic and Maler, 2007). In Pinisetty et al. (2017a), the authors describe how to synthesize an RV monitor for any property defined as a deterministic-timed automata.

The key benefit of online RV monitoring is that the system being monitored is analyzed during its execution. The RV monitoring approaches do not require a formal model of the system, and since only a single execution of the system is considered, these RV monitoring approaches are lightweight and scalable. Thus, in addition to detecting errors, RV techniques can be extended to react whenever an error (incorrect behavior) in a system is detected. The techniques developed for RV are the basis for other techniques that deal with correcting the execution of system at runtime (discussed in 3.3.3).

3.3.3 Correcting Execution of a System at Runtime (Runtime Enforcement)

RE (Schneider, 2000; Ligatti et al., 2009; Falcone et al., 2011; Pinisetty et al., 2012) approaches ensure that a running system satisfies given desired properties. RE approaches can be understood as an extension of RV and refers to the theories, techniques, and tools aimed at guaranteeing the conformance of the execution of the system under observation with respect to some desired property. Using an RE monitor, an untrustworthy input execution in the form of a sequence of events is transformed into an output sequence that complies with the desired property.

As illustrated in Figure 3.7, an enforcement monitor (EM) acts as a filter, whereby it transforms some possibly incorrect execution sequence into a correct sequence with respect to the property of interest. The transformation performed by an EM should satisfy constraints such as *soundness* and *transparency*. Soundness means that the sequence released as output by the EM obeys the property. Transparency means that the monitor should modify the input sequence in a minimal way (e.g. if the input sequence already confirms with the property, then it should not be modified).

The monitors for both RE and RV (shown in Figures 3.7 and 3.6, respectively) take the same data as the input – the property and input events from the system being monitored. They are inserted at the exit (or entrance) of the system, and they do not modify the behavior of the system. However, the goals of these techniques differ. The aim of an RV monitor is to detect misbehaviors or acknowledge desired behavior, which provides the required verdicts for verification. The main aim of RE

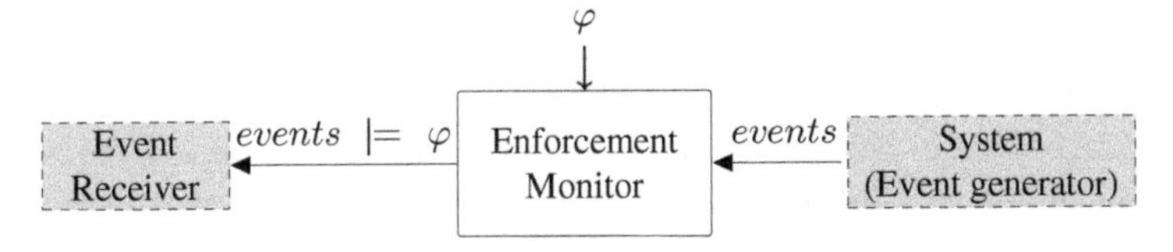

Figure 3.7 Enforcement mechanism.

techniques, however, is to avoid misbehaviors at runtime, and thus, an EM transforms the input event stream into a stream of events that satisfy the desired property.

3.3.3.1 Runtime Enforcement of Untimed Properties

RE has been extensively studied in the context of untimed properties. Several approaches for EMs have been proposed, such as Schneider (2000), Ligatti et al. (2009), Falcone, and Falcone et al. (2011), which differ mainly with the primitives afforded to the monitor – i.e. what the monitor is allowed to do to modify or correct the input sequence.

Security automata (Schneider, 2000) enforce safety properties with a monitor that can block the input sequence. More recently, proposed enforcement monitor mechanisms by Ligatti et al. (2009) enable events from the input sequence to be suppressed (suppression automata), a new event to be inserted (insertion automata), or a new action to either replace the current input or suppress it (edit automata). The set of properties enforced by edit automata is a super-set of safety properties and contains some liveness properties. Similar to edit automata, generic EMs proposed in Falcone et al. (2011) can enforce a set of untimed regular properties.

3.3.3.2 Runtime Enforcement Approaches for Timed Properties

The notion of time has been considered in RE approaches, such as for discrete-time properties (Matteucci, 2007), and the elapse of time as a series of uncontrollable events ("ticks") (Basin et al., 2013).

We now consider some formal approaches for the enforcement of dense time properties such as Pinisetty et al. (2012, 2014a,b,c), Renard et al. (2015), Falcone et al. (2016), Renard et al. (2017), and Renard et al. (2019), which use the expressiveness of timed automaton (TA) (Alur and Dill, 1994). When considering timed properties the input to the EM is a timed word that should be transformed into a timed word that complies with a timed property used to synthesize the EM. In all the frameworks discussed below, EMs are described using denotational paradigms, where EMs are defined as functions through their input/output behavior, and operational paradigms, such as input/output-labelled transition systems and algorithms. These approaches differ in either the supported enforcement operations on the mechanism or the supported classes of properties for which EMs can be generated. The first steps to RE of continuous-timed properties was introduced in Pinisetty et al. (2012), which considers the enforcement of timed safety and cosafety. EMs had only one enforcement operation, which was to delay events, to satisfy the required property. Constraints of the enforcement mechanism ensured that their outputs not only satisfy the required property, but also that the input events are not delayed unnecessarily.

The approach in Pinisetty et al. (2014a,c) generalizes the results presented in Pinisetty et al. (2012) for any regular-timed property. The enforcement operation is still restricted only to the delaying of events. However, it allows for the enforcement of more interesting system properties (e.g. specifying transactional behavior) by considering synthesized EMs for any regular-timed property.

The approach in Falcone et al. (2016) for the enforcement of timed properties considers the enforcement of all regular-timed properties. Regarding possible enforcement operations, in addition to delaying events, the approach in Falcone et al. (2016) also allows suppressing events. Note that priority is given to delaying, and an event is only suppressed if it is not possible to satisfy the property by delaying. The framework proposed in Pinisetty et al. (2014b) considers event-based-timed specifications, where the time between events matters, and also allows events to carry data values from the monitored system. The work introduces the model of parametrized timed automata with variables (PTAVs), which are an extension of TAs with parameters, internal and external variables. It formalizes and defines how parametric-timed specifications (expressed as PTAVs) can be enforced. This is very useful when considering the requirements from practical application domains that have constraints on time and data.

The approach in Renard et al. (2015, 2019) introduces uncontrollable events. The properties to be enforced can feature events that are uncontrollable – ones that can be only observed by the EM and cannot be controlled or acted upon. A framework for enforcing regular-timed properties with uncontrollable events is presented, whereby an EM cannot delay or act upon an uncontrollable event.

When considering reactive systems, terminating the system or delaying the reaction is not a feasible action for an EM. Thus, the approaches discussed so far in this section such as Renard et al. (2015, 2019), Pinisetty et al. (2014b,a,c), and Falcone et al. (2016) are not suitable for reactive systems. In Pinisetty et al. (2017c) an enforcement framework suitable for synchronous reactive systems with bidirectional synchronous EMs is introduced. The synthesis of EMs for properties is expressed using a variant of discrete timed automata (DTAs) (Bozga et al., 1999; Alur and Dill, 1994) is considered. The framework considers similar constraints of soundness and transparency as in the earlier mechanisms. However, the EMs in the framework proposed in Pinisetty et al. (2017c) also satisfies additional requirements of causality and instantaneity, which are specific to synchronous executions.

3.4 Formal Runtime-Based Approaches for Medical Device Security

Here, we will discuss RV as an approach for attack detection. Subsequently, we will discuss RE as a technique for attack mitigation. We illustrate the approaches using a pacemaker case-study.

3.4.1 Overview of the Approach

In this section, we provide a short overview of the heart-pacemaker monitoring device architecture and the approach (shown in Figure 3.8) proposed in Pinisetty et al. (2018).

After a pacemaker is implanted, it remains inside the body for a long duration. If a pacemaker has to be reprogrammed after implantation, it should be done wirelessly

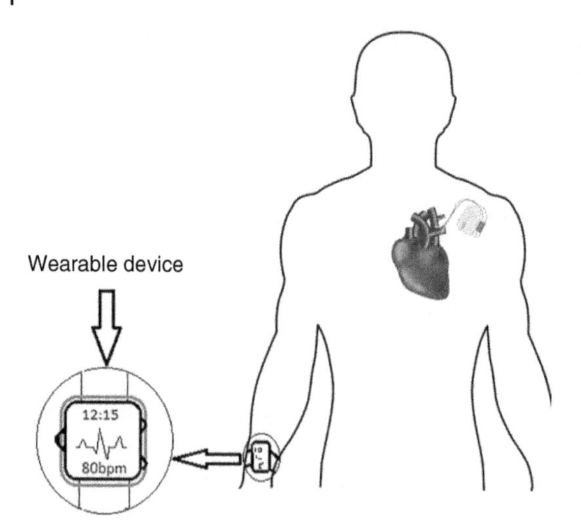

Figure 3.8 Overview of the RV monitoring approach (from Pinisetty et al. (2018)).

through radio frequency transmission. Cardiologists interact with the pacemaker to change operation modes and adjust its timing parameters (Camara et al., 2015). See the discussion in 3.2 for details.

Whenever the pacemaker is implanted, the cardiologist sets the pacemaker sensitivity, the pacing mode (e.g. DDD or DDI), and timers, such as AEI and AVI. When a pacemaker is hacked by an attacker, the attacker may attempt to change (increase/decrease) any of these timers. The proposed approach uses the same timing values on the wearable device as the pacemaker. Also, while the pacemaker has wireless channels, the wearable device uses no wireless connectivity. Thus, the wearable device is assumed to be non-hackable.

The central idea of the developed approach in Pinisetty et al. (2018) is that cardiac electrical activity and the associated timing of the events can be picked up either on the heart surface through the EGM or on the body surface through the ECG. By mapping the EGM-based timing properties to corresponding ECG-based timing properties, the wearable device can deduce if the pacemaker is operating in sync with the originally programmed timing values or not. Through this, the wearable device can deduce the possibility of an attack. In 3.4.2 we provide the mapping of the pacemaker properties to corresponding ECG-properties.

3.4.2 Mapping EGM Properties to ECG Properties

The various key timers for a DDD mode pacemaker and timing intervals for a typical ECG signal were given in 3.2. In Pinisetty et al. (2018) and Pearce et al. (2019b), the authors considered the following set of crucial safety properties from Jiang et al. (2012).

- PM_1: AP and VP should not happen simultaneously.
- PM_2: VS or VP should be true within *AVI* after AS or AP (an atrial event).
- PM_3: AS or AP must be true within *AEI* after VS or VP (a ventricle event).
- PM_4: After a ventricle event, another ventricle event cannot happen before *URI*.
- PM_5: After a ventricle event, within *LRI* another ventricle event must happen.

In the RV monitoring approach proposed in Pinisetty et al. (2018), input event streams for the RV monitor are extracted from an individual's ECGs. It is assumed that the monitoring device has access only to surface ECG and detecting whether properties $PM_1, \cdots, PM_5$ are violated or not is based on feature extraction of ECGs. Thus, the pacemaker timer values such as *AVI* and *AEI* are correlated to the ECG intervals described in 3.2. The pacemaker timers and their corresponding intervals are (Pinisetty et al., 2018):

- The *AVI* timer corresponds to the $P - R$ interval.
- The *AEI* timer corresponds to the time interval beginning from *R wave* till the subsequent *P wave*.
- The *LRI* timer corresponds to the $R - R$ interval.
- The *URI* timer corresponds to the minimum time interval between two consecutive *R waves*.

Using the mapping described above, properties $PM_1, \cdots, PM_5$ can be defined using ECG intervals as follows:

P_1: A *P wave* and *R wave* cannot happen simultaneously.
P_2: *R wave* must arrive within the $P - R$ interval after an *P wave*.
P_3: *P wave* must be true within the $R - P$ interval after an *R wave*.
P_4: After an *R wave*, another *R wave* can come only after the $R - P$ interval.
P_5: After an *R wave*, another *R wave* should come within the $R - R$ interval.

3.4.3 Security of Pacemakers Using Runtime Verification

To verify the critical safety policies defined for heart-pacemaker activity, the authors proposed an externally wearable device using RV techniques that repeatedly monitor the ECG signals of the body, which gives an additional layer of safety as well as security (Pinisetty et al., 2018).

Monitoring solutions for pacemakers, such as Zhang et al. (2013), require communication with the pacemaker, which poses additional security challenges with encryption and key distribution. The alternative is a monitoring mechanism that does not require any communication with the pacemaker. The concept proposed considers that the monitor is deployed on an external wearable device, and relies on ECG sensing to identify events of interest. The cardiologist can program the monitoring device with critical pacemaker timing values. The monitoring device does not require any communication with other devices including the pacemaker, which leads to a reasonable assumption of the robustness and security of the proposed monitoring device (Pinisetty et al., 2018).

The proposed monitor synthesis approach adapts RV for timed automata (TA) (Alur and Dill, 1994) to generate a monitor that identifies anomalous events at runtime. When an anomaly is detected, an alarm alerts the patient. RV is an ideal fit for monitoring the security of IMDs, such as pacemakers, since it is only concerned with runs of the system (treated as a black-box). Thus, no modifications are necessary to the existing pacemaker and additional wireless protocols or associated key distributions are not required. Thus, the device does not require recertification.

We briefly discuss the following topics (Pinisetty et al., 2018):

- A runtime verification framework based on safety policies expressed as timed automata.
- The ECG sensing module for the identification of crucial events. The RV monitor and ECG sensing modules operate together online to verify the desired policies at runtime. An alarm is triggered by the RV monitor whenever a violation of the desired policy is detected.
- The empirical results presented in Pinisetty et al. (2018) illustrate that the overhead of such a monitoring approach is minimal and that such a wearable device can be developed.
- The approach developed in Pinisetty et al. (2018) requires no modifications to be made to pacemakers and their pacing logic. Moreover, no wireless communication modules are required, which minimizes the risk of attacks on the wearable device or the pacemaker using the approach proposed in Pinisetty et al. (2018).

3.4.3.1 Timed Words, Timed Languages, and Defining Timed Properties

In a timed framework, the time instances of action occurrences are also relevant and important. Let $\mathbb{R}_{\geq 0}$ denote the set of non-negative real numbers, and Σ denote a finite set of actions. An event is a pair (t, a), where the action is given by $\text{act}((t,a)) \stackrel{\text{def}}{=} a \in \Sigma$, and the absolute time of the event is given by $\text{date}((t,a)) \stackrel{\text{def}}{=} t \in \mathbb{R}_{\geq 0}$.

A timed word over Σ is a finite sequence of events where $\sigma = (t_1, a_1)\cdot(t_2, a_2)\cdots(t_n, a_n)$, and $(t_i)_{i\in[1,n]}$ is a sequence in $\mathbb{R}_{\geq 0}$ that is nondecreasing. The starting date of σ is denoted using $\text{start}(\sigma) \stackrel{\text{def}}{=} t_1$, and the ending date is denoted with $\text{end}(\sigma) \stackrel{\text{def}}{=} t_n$. The starting and ending dates for empty timed word (ϵ) are null.

Given alphabet Σ, the set of timed words over Σ is denoted by $\text{tw}(\Sigma)$, and any set $\mathcal{L} \subseteq \text{tw}(\Sigma)$ is a timed language. Though the alphabet in timed setting $(\mathbb{R}_{\geq 0} \times \Sigma)$ is infinite, notations in the untimed setting that relate to length, prefix, etc., extend to timed words.

The untimed projection (i.e. ignore dates) of σ is $\Pi_\Sigma(\sigma) \stackrel{\text{def}}{=} a_1 \cdot a_2 \cdots a_n$ in Σ^*. When concatenating two timed words, the dates should be nondecreasing in the resulting timed word. This is ensured if the ending date of the first timed word is less than the starting date of the second timed word. Formally, consider $\sigma = (t_1, a_1)\cdots(t_n, a_n)$ and $\sigma' = (t'_1, a'_1)\cdots(t'_m, a'_m)$ to be two timed words with $\text{end}(\sigma) \leq \text{start}(\sigma')$. Their concatenation is

$$\sigma \cdot \sigma' \stackrel{\text{def}}{=} (t_1, a_1)\cdots(t_n, a_n)\cdot(t'_1, a'_1)\cdots(t'_m, a'_m).$$

By convention $\sigma \cdot \epsilon \stackrel{\text{def}}{=} \epsilon \cdot \sigma \stackrel{\text{def}}{=} \sigma$. Concatenation is undefined otherwise.

Defining Timed Properties In Pinisetty et al. (2018), properties are formally defined as timed automata from which RV monitors are generated. A timed property in Pinisetty et al. (2018) is defined by a timed language that can be recognized by a deterministic and complete TA (Alur and Dill, 1994).

3.4.3.2 Runtime Verification Monitor

Let us now see the definition of the RV monitor for any given timed property φ that is defined as a timed automaton.

Example 3.1 (*TA for property P_4*): *Consider an example of a timed automaton that defines a desired timed property (P_4).*

The TA in Figure 3.9 defines the timed property "After an *R wave*, another *R wave* can come only after the $R - P$ interval" *(property P_4 presented in 3.4.2). The finite set of actions is $\Sigma = \{p, q, r\}$, and the set of finite locations is $L = \{l_0, l_1, l_2\}$, where l_0 is the initial location. The set of real-valued clocks is $X = \{x\}$. As illustrated in Figure 3.9, guards with constraints on clock values are present on transitions, e.g. $x < RP$ on the transition between l_1 and l_2 (where $RP \in \mathbb{N}$) and clock resets. When the first r action occurs, the automaton moves to l_1 from l_0, and the clock x is reset to 0. When in location l_1, if action r occurs and if $x \geq RP$, then the automaton remains in l_1, and the value of clock x is reset to 0. It moves to location l_2 otherwise. The nonaccepting location l_2 should never be reached for the property to be satisfied over runs.*

Definition 3.1 (*RV monitor*): Consider $\varphi \subseteq \text{tw}(\Sigma)$ (represented as TA $\mathcal{A}_\varphi$), which defines the property that the monitor is given. Function $M_\varphi : \text{tw}(\Sigma) \rightarrow \mathcal{D}$ is a verification monitor for φ, where $\mathcal{D} = \{\mathsf{true}, \mathsf{false}, \mathsf{c\_true}, \mathsf{c\_false}\}$, with $\sigma \in \text{tw}(\Sigma)$ representing the current observation. Function M_φ is defined as follows:

$$M_\varphi(\sigma) = \begin{cases} \mathsf{true} & \text{if } \forall \sigma' \in \text{tw}(\Sigma) : \sigma \cdot \sigma' \in \varphi \\ \mathsf{false} & \text{if } \forall \sigma' \in \text{tw}(\Sigma) : \sigma \cdot \sigma' \notin \varphi \\ \mathsf{c\_true} & \text{if } \sigma \in \varphi \wedge \exists \sigma' \in \text{tw}(\Sigma) : \sigma \cdot \sigma' \notin \varphi \\ \mathsf{c\_false} & \text{if } \sigma \notin \varphi \wedge \exists \sigma' \in \text{tw}(\Sigma) : \sigma \cdot \sigma' \in \varphi. \end{cases}$$

In Definition 3.1, verdicts *true* (true) and *false* (false) are conclusive, and *currently true* (c_true), and *currently false* (c_false) are inconclusive. An inconclusive verdict states an evaluation about the current observation (execution seen so far).

$M_\varphi(\sigma)$ returns conclusive verdict true if for any continuation $\sigma' \in \text{tw}(\Sigma)$, $\sigma \cdot \sigma'$ *satisfies* φ. If for any continuation $\sigma' \in \text{tw}(\Sigma)$, $\sigma \cdot \sigma'$ *falsifies* φ then $M_\varphi(\sigma)$ returns conclusive verdict false.

Monitor $M_\varphi(\sigma)$ returns inconclusive verdict c_true if the current observation σ satisfies φ, and if there is a continuation $\sigma' \in \text{tw}(\Sigma)$ s.t. $\sigma \cdot \sigma'$ does not satisfy φ (i.e.

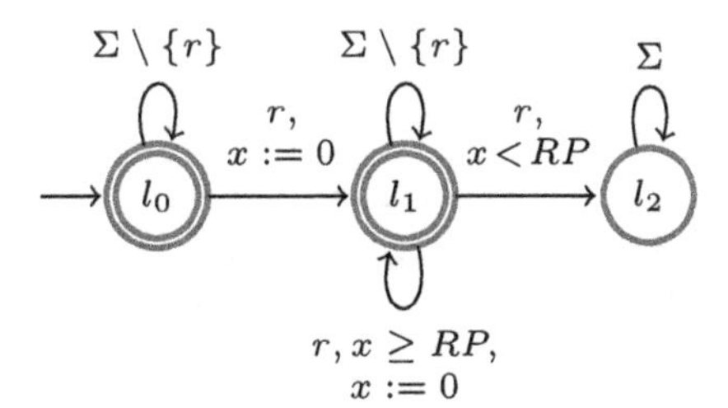

Figure 3.9 Timed automaton defining property P_4 in 3.4.2 (from Pinisetty et al. (2018)).

not all continuations of σ satisfy φ). $M_\varphi(\sigma)$ returns inconclusive verdict c_false if σ does not satisfy φ, and there is a continuation $\sigma' \in \text{tw}(\Sigma)$ such that $\sigma \cdot \sigma'$ satisfies φ.

Example 3.2 (*Example illustrating the behavior of an RV monitor*): *Consider monitoring property P_4 from 3.4.2, defined by the TA illustrated in Figure 3.9. Let RP be 850 time units. Table 3.1 presents how the monitor for P_4 behaves when the input timed word $\sigma = (30, p) \cdot (228, r) \cdot (320, p) \cdot (471, r)$ is processed step-by-step.*

At t $= 30$, when the current observed input is $\sigma = (30, p)$, σ satisfies P_4 but there are some extensions $\sigma' \in tw(\Sigma)$ such that $\sigma \cdot \sigma'$ does not satisfy the property P_4. So, the monitor provides the verdict c_true *in the first step. Similarly, the monitor provides the verdict* c_true *in the next two steps at* t $= 228$ *and* t $= 320$. *At* t $= 471$, *after observing the event* $(471, r)$ *(i.e. when the current observed input is* $\sigma = (30, p) \cdot (228, r) \cdot (320, p) \cdot (471, r)$*), the property φ is falsified by σ and for any of its extension (i.e., for any $\sigma' \in tw(\Sigma)$, $\sigma \cdot \sigma'$ falsifies the property φ). Thus, immediately after observing* $(471, r)$ *the monitor provides a conclusive verdict (false).*

External Wearable Device In the approach proposed in Pinisetty et al. (2018), it is assumed the patient will also wear a monitoring device after the pacemaker is implanted. As illustrated in Figure 3.8, the external wearable monitoring device could be a smart watch or any computing device with an ECG sensor and an accelerometer. When configuring the pacemaker, the doctor also configures the external wearable device with timing values, whereby all the timer values are stored in the wearable device's memory. It also knows the normal heart rate at which the pacemaker is set to pace (e.g. 60 to 120 bpm). To monitor the activity of the body, the device may also have a built-in accelerometer.

The monitoring device records the surface ECG signals via the ECG sensor. It can extract all the relevant actions of interest from the ECG signals (the peaks of P, Q, R, S, and T waves as well as the pacing pulses) by filtering and processing the ECG signal data. Using RV monitors, the violation of any safety properties can be checked depending on the time instances at which different actions (peaks) occur. If a violation of any desired property is observed, the monitoring device can generate an alarm for the user. As illustrated in Figure 3.8, the monitoring device does not require any direct communication with the pacemaker.

Table 3.1 Example behavior of an RV monitor for property P_4.

σ	$M_\varphi(\sigma)$
$(30, p)$	c_true
$(30, p) \cdot (228, r)$	c_true
$(30, p) \cdot (228, r) \cdot (320, p)$	c_true
$(30, p) \cdot (228, r) \cdot (320, p) \cdot (471, r)$	false

3.4.3.3 Architecture of the Monitoring System

The monitoring system consists of two modules as illustrated in Figure 3.10: (i) `ECG_Processing` module and (ii) `RV_Monitor` module. The modules run concurrently in an *on-line* manner. The `ECG_Processing` module analyses ECGs (performs real-time signal processing) to detect the events of interest which are passed as timed events to the online `RV_Monitor` module.

`ECG_Processing` module: For monitoring the desired properties ($P_1, \cdots, P_5$ introduced in 3.4.2), all the relevant events such as P, Q, and R peaks should be extracted from ECG data. This module filters and processes the ECG data to detect these actions (which, along with the appropriate timing information, are fed as input events to the `RV_Monitor` module).

`RV_Monitor` module: This module takes as its input the property to be verified (φ) and a stream of timed events (or an event stream generated by the ECG processing module), and determines if the input event stream satisfies (or violates) φ.

Remark 3.1 The properties considered ($P_1, \cdots, P_5$) are timed safety properties (which can be defined as timed automata (Alur and Dill, 1994)), for which RV monitors can be synthesized using approaches such as Pinisetty et al. (2017b). However, verifying these properties implicitly detects attacks as follows. As discussed earlier, an attacker may gain access to the pacemaker device and change the programmed timing values, which may lead to serious harm to the patient. However, the proposed wearable monitoring device cannot be programmed wirelessly and is secure (e.g. using authentication mechanisms such as human biometrics). Thus, the attacker is unable to access the wearable monitoring device. When the pacemaker device is attacked, the monitoring device will detect it within a short time due to the mismatch in timing values that lead to violation of the considered safety properties.

3.4.3.4 Implementation of the ECG Processing and RV Monitor Modules

ECG Processing Module To demonstrate the feasibility of the proposed approach, in Pinisetty et al. (2018) prerecorded ECG data with pacing artifacts were used. However, by feeding ECG cycles one-by-one for processing, the data are used as if they were generated in real time. The function of the `ECG_Processing` module and a description of how ECG signals are processed to extract all relevant actions of interest for monitoring the considered set of properties is discussed in Pinisetty et al. (2018). An implementation of `ECG_Processing` in MATLAB and the data set from Phy considered for experiments is also described in detail in Pinisetty et al. (2018).

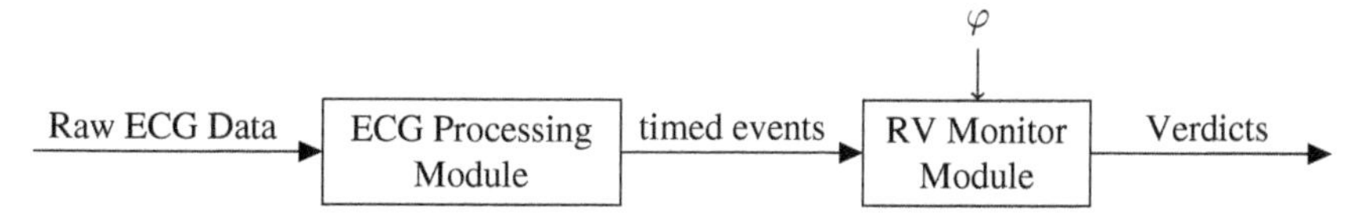

Figure 3.10 Architecture of the RV monitor (from Pinisetty et al. (2018)).

Online RV Monitor Module Earlier in this section, we briefly discussed the functional description of a RV monitor for a given property φ. An online RV-monitoring algorithm (taking the TA defining property to monitor as an input parameter) based on the functional description is detailed in Pinisetty et al. (2018). The algorithm is implemented in Python using UPPAAL (dbm) libraries.

3.4.3.5 Summary of Experiments and Results

To demonstrate the practicality of the approach, a prototype was developed that shows how the monitor reacts when it receives a new event and identifies violation(s) of the properties of interest. The prototype consists of modules such as the `ECG_Processing` module and the `RV_Monitor` module.

The prototype demonstrates how the monitor identifies violation(s) of the properties of interest and reacts when it receives events one-by-one. Details of the implementation and experiments illustrating the input/output behavior of the monitor via some examples are described in Pinisetty et al. (2018). Experiments were also conducted to evaluate the effect of RV algorithms on the execution time. Based on the monitoring execution time overhead, a monitoring device using the proposed approach will be sufficiently fast to check for any anomalies in heart activity (Pinisetty et al., 2018).

3.4.4 Securing Pacemakers with Runtime Enforcement Hardware

In 3.4.3, we briefly discussed formal RV monitoring as a methodology for monitoring the status of medical devices. This is important, especially within the context of modern pacemaker devices, which are increasingly providing greater and greater attack surfaces for malicious actors. Wireless control features, while extremely helpful for doctors and medical staff, also enable new kinds of attacks. In certain cases, these attacks can be performed beyond the immediate vicinity of the patient, with internet-enabled devices potentially vulnerable to attacks that originate anywhere in the world (P. and Aziz, 2018). Further, while tools such as external monitors can be used to *detect* attacks (for instance the RV monitor could be used to generate an alert or warning), there will be a delay between the alert and the intervention triggered by that alert. As such, there is still scope for the attack to succeed (for instance, a patient may be harmed before a medical professional is able to resolve the situation).

Traditional methodologies for preventing malicious attacks in cyber-enabled systems do not always apply well to battery-operated devices, such as pacemakers and other IMDs. Even simple tools such as encryption may not be included due to constraints on processing power and battery life (in one example, an insulin sensor had to operate on a 1.5 volt watch battery for two years (Takahashi, 2011)). In addition, even where these tools are used, they still feature weaknesses (Marin et al., 2016) and there is much evidence in practice that it is both infeasible and impractical to eliminate all security vulnerabilities for any given system (Sametinger et al., 2015).

In this section, we briefly discuss a solution to this problem through *runtime enforcement*. This approach was first proposed in Pearce et al. (2019b) for ensuring

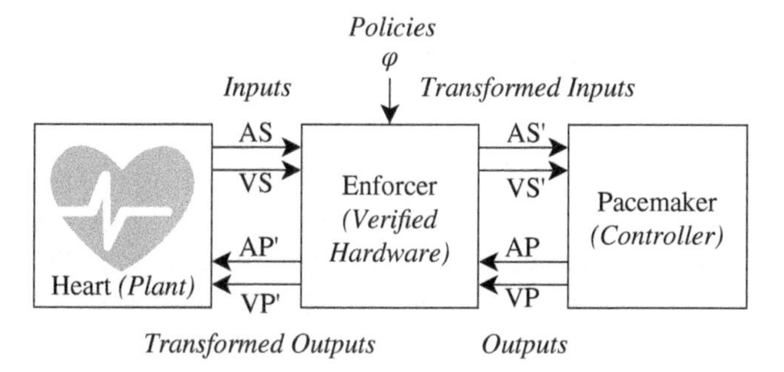

Figure 3.11 Pacemaker with runtime enforcer (from Pearce et al. (2019b)).

the safety of patients with IMDs *even in the presence of a malicious attack*. The authors illustrate the approach via a pacemaker running example. Here, REs are external hardware that are computationally isolated from the medical device controllers and are designed such that a given IMD will continue to operate at a *baseline level of safety* even in the presence of a malicious attack. Figure 3.11 provides an overview of the approach. As illustrated, the system is composed of an enforcer located between the plant (e.g. the heart) and the controller (e.g. the pacemaker). The hardware's timing and functional properties are further verified to ensure correct operation, and power consumption requirements are also estimated.

3.4.4.1 Preliminaries: Words, Languages, and Defining Properties as DTA

Discrete timed automata (DTA) are timed automata with a set of integer-valued clocks (Bozga et al., 1999). The discrete integer clocks are, for instance, used to count the number of ticks before a certain event occurs. The complexity of TA-based enforcement is shown to be PSPACE-complete due to the need for reachability computation (Pinisetty, 2015). The RE approach may be more scalable considering DTA (Pinisetty et al., 2017c), and considering integer-valued clocks is more suitable for the approach in Pearce et al. (2019b), which considers the synthesis of hardware enforcers.

To formally define policies for reactive systems that use finite, ordered sets of Boolean signals as inputs $I = \{i_1, i_2, \cdots, i_n\}$ and outputs $O = \{o_1, o_2, \cdots, o_n\}$, DTA (Pinisetty et al., 2017c) are more suitable and used in Pearce et al. (2019b). Consider the input alphabet $\Sigma_I = 2^I$ when the output alphabet is $\Sigma_O = 2^O$ and the input/output alphabet is $\Sigma = \Sigma_I \times \Sigma_O$. An input/output event a is of the form $a = (x, y)$, where $x \in \Sigma_I$ and $y \in \Sigma_O$.

Example 3.3 (*Input/output alphabet: Pacemaker I/O*): *Let us consider the pacemaker example scenario described in 3.2.1, where $I = \{AS, VS\}$ and $O = \{AP, VP\}$. The input alphabet is $\Sigma_I = \{00, 01, 10, 11\}$, where for each element in Σ_I, the first bit in represents AS and the second represents VS. Similarly the output alphabet is $\Sigma_O = \{00, 01, 10, 11\}$, and $\Sigma = \{(00,00), (00,01), \cdots, (11,10), (11,11)\}$ is the input/output alphabet.*

A finite (infinite) word from any alphabet Σ is a finite sequence $\sigma = a_1 \cdot a_2 \cdots a_n$ (or infinite sequence $\sigma = a_1 \cdot a_2 \cdots$) of elements of Σ. The set of finite (infinite) words over Σ is denoted by Σ^* (or Σ^ω).

Defining Timed Properties In Pearce et al. (2019b), properties are formally defined as DTA (Pinisetty et al., 2017c), from which RE monitors are generated. The properties related to the pacemaker case study considered are (P_1 through P_5) discussed earlier in this section. The DTA can be defined for each property and combined using the product of DTAs (Pinisetty et al., 2017c).

Remark 3.2 The properties related to the pacemaker case study considered are (P_1 through P_5) discussed earlier in this section. Note that while the timing values are typically real-valued constants, they are discretized in Pearce et al. (2019b) based on some digital clock frequency f. For example, consider C a real-value, it can be approximated to C_{cycles} by using either $\lceil \frac{C}{f} \rceil$ or $\lfloor \frac{C}{f} \rfloor$.

3.4.4.2 Runtime Enforcement Monitor

A policy φ over some alphabet Σ allow us to formally define *correct* and *incorrect* behavior. In the work proposed in Pearce et al. (2019b), properties to be enforced are specified formally as discrete timed automaton (DTA) (Pinisetty et al., 2017c). From the policy defined as discrete timed automaton (DTA), an enforcement monitor is synthesized.

Example 3.4 (*Policy P_4 DTA*): *Let us consider the pacemaker policy P_4 discussed earlier. It can be represented as the DTA $\mathcal{A}_{P_4}$, presented in Figure 3.12. Let us recall that the policy informally states that after any VS or VP event, no VP event can be released for at least URI_{cycles}. The set of locations in the DTA is $L = \{l_0, l_1, l_2\}$, where l_0 is the initial location. When in l_0, the DTA waits for the input VS or VP, and once received, the transition to l_1 resets clock v to zero. The set of discrete clocks contains v, which is responsible for counting the passage of time, and increments once every cycle (i.e. by one for every invocation of the DTA). When $v \geq URI_{cycles}$, then the transition to l_0 is taken, indicating that the wait period is finished. In case an VP event occurs before that*

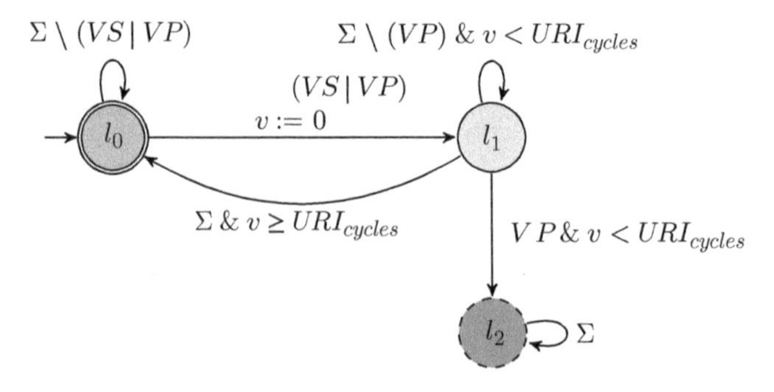

Figure 3.12 Simplified DTA for policy P_4, $\mathcal{A}_{P_4}$ (from Pearce et al. (2019b)).

time has elapsed, then the transition to l_2 will be taken instead. In this DTA, l_2 is the non-accepting trap location that should never be reached.

Synthesis of Runtime Enforcers from Policies Specified as DTA (Edit Functions) A policy DTA can identify nonaccepting I/O traces at runtime. RE mechanisms use the policy DTA to also identify when an I/O signal should be *edited* to avoid violation of the policy at runtime. For example, to prevent the transition to the violation state l_2, an enforcer for policy $\mathcal{A}_{P_4}$ will discard any additional VP signals after receiving a VS or VP signal while the timer v is less than URI_{cycles}.

In Pearce et al. (2019b), the edit behavior is formalized by the edit functions. As illustrated in Figure 3.11, enforcers can operate bidirectionally, i.e. over both inputs and outputs in the general case. To achieve bidirectional enforcement, enforcers operate by first transforming the inputs (from the plant to the controller), and then transforming the outputs (from the controller to the plant). To perform enforcement over the inputs, the policy DTA $\mathcal{A}$, and its alphabet is projected over the inputs (Pearce et al., 2019b), called the input DTA. The input-edit function is defined using the input DTA, and the original DTA is used for defining the output-edit function (Pearce et al., 2019b).

For a given state q_I in the input DTA, the input-edit function will return all possible input events that can be extended by any input trace that leads to an accepting state. The output-edit function, using the original DTA, defines that for the given state q and the input x, all possible input-output events $(x, y) \in \Sigma$ that can be extended by an input-output trace σ' are returned such that σ' is accepting. Definitions and explanation of the edit functions are given in Pearce et al. (2019b).

Synthesis of RE Hardware The approach proposed in Pearce et al. (2019b) considers that the plant, the enforcer, and *interface registers* are synchronously composed together. As illustrated in Figure 3.13, the interface registers decouple the enforcer and the controller, and the interface registers encapsulate the controller as a new black-box controller.

The generalized hardware enforcer is illustrated in Figure 3.14. Enforcers operate iteratively; inputs are first processed (and edited, if necessary), and then outputs are processed (and edited, if necessary). Details regarding the construction of each component are provided in Pearce et al. (2019b). The formal theory and algorithms

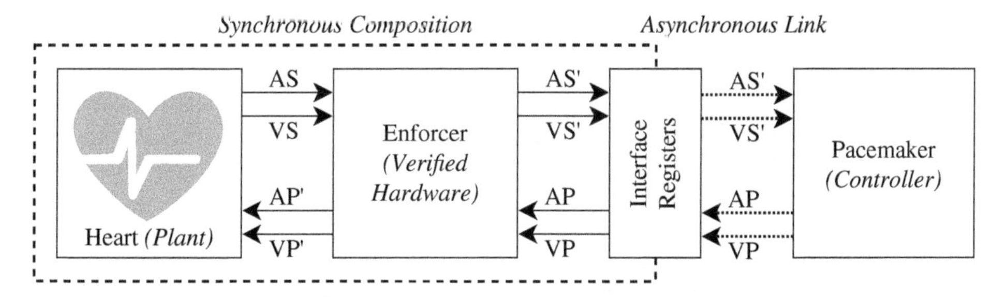

Figure 3.13 System composition (from Pearce et al. (2019b)).

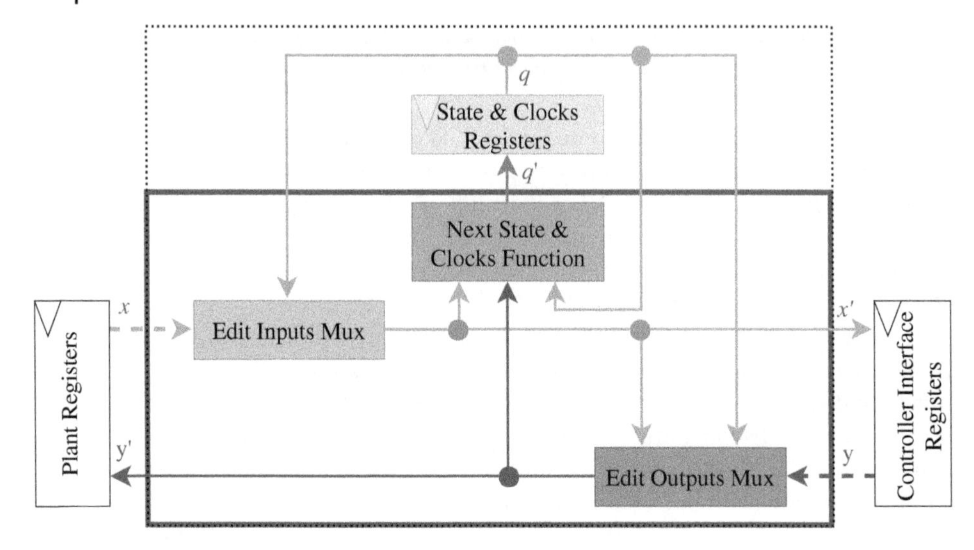

Figure 3.14 Generalized enforcer hardware (from Pearce et al. (2019b)).

for the synthesis of enforcers from policies defined as DTA are detailed in Pinisetty et al. (2017c). The semantics of DTA are encapsulated in hardware in Pearce et al. (2019b) as a mealy-type finite state machine.

3.4.4.3 Verification of the Enforcer Hardware

EBM is a model checker suitable for the verification of hardware designs. REs have complete control of a system's I/O signals. Thus, they could lead to catastrophic consequences in the case of design defects. Though the formal approach of the synthesis of enforcers from policies defined as DTA guarantee that the synthesized enforcers satisfy constraints (such as Soundness, Monotonicity, Instantaneity, Transparency, and Causality) (Pinisetty et al., 2017c), it is not certain that a specific *implementation* of those semantics will be correct. Thus, in Pearce et al. (2019b), the authors propose the use of a *hardware model checker* to verify that the implementation of the design matches the requirements of the system. Using EBMC, certain desired properties of the generated enforcer (compiled Verilog) are verified. Further details are provided in Pearce et al. (2019b).

3.4.4.4 How Does the Enforcer Prevent Security Attacks?

Let us now discuss briefly how the enforcer for the policies considered ($P_1, \cdots, P_5$) prevents security attacks. As discussed, policies $P_1, \cdots, P_5$ are safety policies that govern the correct safe behavior for our pacemaker system. If the pacemaker fails to fulfil these policies, it could be due a successful attack. For example, attack scenarios could be:

1. Attacker turns off the pacemaker pacing function.
2. Attacker reprograms the control software to pace too quickly.
3. Attacker reprograms the control software to trigger AP and VP simultaneously.

Situations such as the above are effectively mitigated by the enforcer for policies $P_1, \cdots, P_5$. In the case when the attacker turns off the pacing function, P_2, P_4, and P_5 would cause the enforcer to provide pace commands in place of the pacemaker software. Similarly, if the attacker reprograms the pacemaker to pace too quickly, the enforcer for P_4 would suppress some of the pacing events. The enforcer for policy P_1 prevents AP and VP from occurring simultaneously.

3.4.4.5 Summary of Experiments and Results

In Pinisetty et al. (2018), results were collected and analyzed using the pacemaker properties P_1 through P_5 discussed earlier in this section. After defining the policies as DTA using an XML-based format, the policies were compiled to Verilog using the approach described in Pearce et al. (2019b). Then, using the EBMC model-checker, each enforcer was verified. The Verilog was later synthesized by using Intel Quartus 16.0 targeting a Max V 5M570ZT100C5 complex programmable logic device (CPLD) (opted for in Pearce et al. (2019b) due to their low power requirements). Quartus Powerplay was used to estimate power for the CPLD.

Detailed benchmarking and analysis results considering different combinations of policies P_1 through P_5 are provided in Pearce et al. (2019b). From the analysis results, it is observed that with an increase in the complexity (the number of states, transitions, and clocks) of the policy, the complexity of the enforcer that is generated also increases. This is evident in the analysis that observed an increasing time required for verification by EBMC, and an increasing number of logic elements (LEs) to represent the enforcer. However, the results reported in Pearce et al. (2019b) indicated that all benchmarks operated considerably faster than the average speed of their software counterparts. It is also observed via the benchmarks that the approach is scalable, as the hardware requirements of any enforcer are linearly related to the state space of the DTA.

3.5 Summary

CP attacks on many critical infrastructure such as smart grids, nuclear power plants, gas pipelines, industrial automation systems, and intelligent transportation systems pose serious risks. These are examples of CPS where the attacker can risk the safety of physical infrastructure with the potential for causing serious harm. Therefore, there have been many studies on attack detection and mitigation of CP attacks (Chen et al., 2017; Lanotte et al., 2018). As these systems are safety-critical in nature, there has been a focus on techniques inspired by formal methods (Lanotte et al., 2017).

In this chapter, we concentrated on formal techniques for attack detection and mitigation of CP attacks on medical devices. We focused on the pacemaker example, which is used for treating bradycardia. The pacemaker was selected as it is an excellent example of a CPS, where it acts as a controller to control the operations of the heart. Moreover, as pacemakers also adopt the modern trend of increasing complexity with new enhancements and wireless features, they become increasingly difficult to comprehensively verify and test, especially in the presence of malicious third parties. With greater feature sets come greater attack surfaces, and as reported by certification agencies such as the FDA, adversaries have already begun exploiting these (Weaver, 2013). As these attacks have the potential to cause serious harm, including death, there is widespread interest on the topic of securing such devices (Camara et al., 2015).

In this chapter, we provided a systematic overview of the problem. We also provided an overview of the bio-physical signals involved and how they could be used in attack detection and mitigation. In the process, we have provided an overview of pacemaker heart interactions in a closed-loop and the associated timing properties, which are essential for safe operation. We introduced general formal methods, and detailed two different formal techniques for attack detection and attack mitigation respectively. However, some avenues remain unexplored – while the framework presented in this chapter is ideal for securing local I/O (such as the leads of a pacemaker), the distribution of that I/O over larger systems remains an open problem. Further, the current framework is designed to work with simple binary control signals (on/off) as it is intended as being as the last layer between the controller and the physical actuators. It would be interesting to examine the expansion of it to support more comprehensive communication protocols between plant and controller so that the framework could be included at higher levels of the design.

References

EBMC. http://www.cprover.org/ebmc. (accessed 20 December 2020).

MetaMath. http://us.metamath.org/. (accessed 20 December 2020).

PhysioNet. http://physionet.org/. (accessed 14 May 2018).

Uppaal DBM Library. http://people.cs.aau.dk/adavid/UDBM/. (accessed 14 May 2018).

H. Alemzadeh, R.K. Iyer, Z. Kalbarczyk, and J. Raman. Analysis of safety-critical computer failures in medical devices. *Security & Privacy, IEEE*, 11(4):14–26, 2013.

R. AlTawy and A.M. Youssef. Security tradeoffs in cyber physical systems: A case study survey on implantable medical devices. *IEEE Access*, 4:959–979, 2016.

R. Alur. *Principles of cyber-physical systems*. MIT Press, 2015.

R. Alur and D.L. Dill. A theory of timed automata. *Theoretical Computer Science*, 126:183–235, 1994. doi/10.1016/0304-3975(94)90010-8.

C. Baier and J.P. Katoen. *Principles of model checking*. MIT Press, 2008. ISBN 978-0-262-02649-9.

H. Barringer, Y. Falcone, K. Havelund, G. Reger, and D.E. Rydeheard. Quantified event automata: Towards expressive and efficient runtime monitors. In *FM*, pages 68–84, 2012.

D. Basin, F. Klaedtke, and E. Zalinescu. Algorithms for monitoring real-time properties. In *Proceedings of the 2nd International Conference on Runtime Verification (RV 2011), volume 7186 of Lecture Notes in Computer Science*, pages 260–275. Springer-Verlag, 2011. ISBN 978-3-642-29859-2. doi/10.1007/978-3-642-29860-8_20.

D. Basin, V. Jugé, F. Klaedtke, and E. Zălinescu. Enforceable security policies revisited. *ACM Transactions on Information and System Security*, 16(1):3:1–3:26, June 2013. ISSN 1094-9224. doi/10.1145/2487222.2487225.

A. Bauer, M. Leucker, and C. Schallhart. Runtime verification for LTL and TLTL. *ACM Transactions on Software Engineering and Methodology*, 20(4):14:1–14:64, September 2011. ISSN 1049-331X. doi/10.1145/2000799.2000800.

Y. Bertot and P. Castéran. *Interactive Theorem Proving and Program Development. Coq'Art: The Calculus of Inductive Constructions*. Springer, 2004. URL http://www.labri.fr/perso/casteran/CoqArt/index.html. accessed 20 December 2020.

Z.E. Bhatti, P.S. Roop, and R. Sinha. Unified functional safety assessment of industrial automation systems. *IEEE Transactions on Industrial Informatics*, 13(1):17–26, 2016.

H. Bohnenkamp and A. Belinfante. Timed testing with torx. In J. Fitzgerald, I.J. Hayes, and A. Tarlecki, editors, *FM 2005: Formal Methods, volume 3582 of Lecture Notes in Computer Science*, pages 173–188. Springer Berlin Heidelberg, 2005. ISBN 978-3-540-27882-5. doi/10.1007/11526841_13.

M. Bozga, O. Maler, and S. Tripakis. Efficient verification of timed automata using dense and discrete time semantics. In L. Pierre and T. Kropf, editors, *Correct Hardware Design and Verification Methods*, pages 125–141, Berlin, Heidelberg, 1999. Springer Berlin Heidelberg. ISBN 978-3-540-48153-9.

C. Camara, P. Peris-Lopez, and J.E. Tapiador. Security and privacy issues in implantable medical devices: A comprehensive survey. *Journal of Biomedical Informatics*, 55:272–289, 2015.

M.R.V. Chaudron, W. Heijstek, and A. Nugroho. How effective is uml modeling? *Software & Systems Modeling*, 11(4):571–580, 2012.

F. Chen and G. Rosu. Parametric trace slicing and monitoring. In *15th International Conference on Tools and Algorithms for the Construction and Analysis of Systems*, volume 5505 of *LNCS*, pages 246–261, 2009. ISBN 978-3-642-00767-5.

T.M. Chen. Stuxnet, the real start of cyber warfare?[editor's note]. *IEEE Network*, 24(6):2–3, 2010.

Y. Chen, S. Kar, and J.M.F. Moura. Cyber-physical attacks with control objectives. *IEEE Transactions on Automatic Control*, 63(5):1418–1425, 2017.

D. Clarke, T. Jéron, V. Rusu, and E. Zinovieva. STG: A symbolic test generation tool. In J.P. Katoen and P. Stevens, editors, *Tools and Algorithms for the Construction and Analysis of Systems*, volume 2280 of *Lecture Notes in Computer Science*, pages 470–475. Springer Berlin Heidelberg, 2002. ISBN 978-3-540-43419-1. doi/10.1007/3-540-46002-0_34.

D. Clery. Could your pacemaker be hackable? *Science*, 347(6221):499–499, 2015. ISSN 0036-8075. doi/10.1126/science.347.6221.499.

C. Colombo, G.J. Pace, and G. Schneider. LARVA — safer monitoring of real-time Java programs (tool paper). In *Proceedings of the 7th IEEE International Conference on Software Engineering and Formal Methods (SEFM 2009)*, pages 33–37. IEEE Computer Society, 2009. ISBN 978-0-7695-3870-9. doi/10.1109/SEFM.2009.13.

V. D'Silva, D. Kroening, and G. Weissenbacher. A survey of automated techniques for formal software verification. *IEEE Trans. on CAD of Integrated Circuits and Systems*, 27(7):1165–1178, 2008. doi/10.1109/TCAD.2008.923410.

D. Evans and D. Larochelle. Improving security using extensible lightweight static analysis. *IEEE Software*, 19(1):42–51, January 2002. ISSN 0740-7459. http://dx.doi.org/10.1109/52.976940.

Y. Falcone. You should better enforce than verify. In: *Proceedings of the 1st international conference on Runtime verification (RV 2010)*, Lecture Notes in Computer Science, vol. 6418, pp. 89–105, Springer-Verlag, 2010.

Y. Falcone and L.D. Zuck. Runtime verification: The application perspective. In T. Margaria and B. Steffen, editors, *Leveraging Applications of Formal Methods, Verification and Validation. Technologies for Mastering Change*, volume 7609 of *Lecture Notes in Computer Science*, pages 284–291. Springer Berlin Heidelberg, 2012. ISBN 978-3-642-34025-3. doi/10.1007/978-3-642-34026-0_21.

Y. Falcone, L. Mounier, J.C. Fernandez, and J.L. Richier. Runtime enforcement monitors: composition, synthesis, and enforcement abilities. *Formal Methods in System Design*, 38(3):223–262, 2011.

Y. Falcone, T. Jéron, H. Marchand, and S. Pinisetty. Runtime enforcement of regular timed properties by suppressing and delaying events. *Science of Computer Programming*, 123:2–41, 2016. doi/10.1016/j.scico.2016.02.008.

K. Forsberg and H. Mooz. The relationship of system engineering to the project cycle. *INCOSE International Symposium*, 1(1):57–65, 1991. doi/10.1002/j.2334-5837.1991.tb01484.x.

H. He and J. Yan. Cyber-physical attacks and defences in the smart grid: a survey. *IET Cyber-Physical Systems: Theory & Applications*, 1(1):13–27, 2016.

G.J. Holzmann. The model checker spin. *IEEE Transactions on Software Engineering*, 23(5):279–295, May 1997. ISSN 0098-5589. doi/10.1109/32.588521.

L. Hubert, N. Barré, F. Besson, D. Demange, T. Jensen, and et al. Sawja: Static analysis workshop for java. In B. Beckert and C. Marché, editors, *Formal Verification of Object-Oriented Software*, volume 6528 of *Lecture Notes in Computer Science*, pages 92–106. Springer Berlin Heidelberg, 2011. ISBN 978-3-642-18069-9. doi/10.1007/978-3-642-18070-5_7.

M. Huth and M. Ryan. *Logic in Computer Science: Modelling and Reasoning About Systems*. Cambridge University Press, New York, NY, USA, 2004. ISBN 052154310X.

T. Jan. Testing techniques. *Lecture Notes*, 2002. URL http://www-i2.informatik.rwth-aachen.de/dl/mbt08/lec_notes_04.pdf.

C. Jard and T. Jéron. TGV: Theory, principles and algorithms: A tool for the automatic synthesis of conformance test cases for non-deterministic reactive systems. *International Journal on Software Tools for Technology Transfer*, 7(4):297–315, August 2005. ISSN 1433-2779. doi/10.1007/s10009-004-0153-x.

Z. Jiang, M. Pajic, S. Moarref, R. Alur, and R. Mangharam. Modeling and verification of a dual chamber implantable pacemaker. In *TACAS*, TACAS'12, pages 188–203, Berlin, Heidelberg, 2012. Springer-Verlag. ISBN 978-3-642-28755-8. doi/10.1007/978-3-642-28756-5_14.

K.B. Kelarestaghi, K. Heaslip, M. Khalilikhah, A. Fuentes, and V. Fessmann. Intelligent transportation system security: hacked message signs. *SAE International Journal of Transportation Cybersecurity and Privacy*, 1 (11-01-02-0004):75–90, 2018.

J. Kirk. Pacemaker hack can deliver deadly 830-volt jolt. *Computerworld*, 17, 2012. (accessed day month, year).

M. Krichen and S. Tripakis. Conformance testing for real-time systems. *Formal Methods in System Design*, 34(3):238–304, June 2009. ISSN 0925-9856. doi/10.1007/s10703-009-0065-1.

R. Lanotte, M. Merro, R. Muradore, and L. Viganò. A formal approach to cyber-physical attacks. In *2017 IEEE 30th Computer Security Foundations Symposium (CSF)*, pages 436–450. IEEE, 2017.

R. Lanotte, M. Merro, and S. Tini. Towards a formal notion of impact metric for cyber-physical attacks. In C.A. Furia and K. Winter, editors, *Integrated Formal Methods*, pages 296–315, Cham, 2018. Springer International Publishing. ISBN 978-3-319-98938-9.

K.G. Larsen, P. Pettersson, and W. Yi. UPPAAL in a nutshell. *International Journal on Software Tools for Technology Transfer*, 1:134–152, 1997.

J. Ligatti, L. Bauer, and D. Walker. Run-time enforcement of nonsafety policies. *ACM Transactions on Information and System Security*, 12(3):19:1–19:41, January 2009. ISSN 1094-9224. doi/10.1145/1455526.1455532.

G. Loukas. *Cyber-physical attacks: A growing invisible threat*. Butterworth-Heinemann, 2015.

S. Madnick and M.E. Mangelsdorf. What executives get wrong about cybersecurity. *Sloan Management Review*, pages 22–24, 2017.

W.H. Maisel, M.O. Sweeney, W.G. Stevenson, K.E. Ellison, and L.M. Epstein. Recalls and Safety Alerts Involving Pacemakers and Implantable Cardioverter-Defibrillator Generators. *JAMA*, 286(7):793–799, 08 2001. ISSN 0098-7484. doi/10.1001/jama.286.7.793.

E. Marin, D. Singelée, F.D. Garcia, T. Chothia, R. Willems, and B. Preneel. On the (in) security of the latest generation implantable cardiac defibrillators and how to secure them. In *Proceedings of the 32nd annual conference on computer security applications*, pages 226–236, 2016.

I. Matteucci. Automated synthesis of enforcing mechanisms for security properties in a timed setting. *Electronic Notes in Theoretical Computer Science*, 186:101–120, July 2007. ISSN 1571-0661. doi/10.1016/j.entcs.2007.03.025.

H.G. Mond and A. Proclemer. The 11th world survey of cardiac pacing and implantable cardioverter-defibrillators: calendar year 2009–a world society of arrhythmia's project. *Pacing and Clinical Electrophysiology : PACE*, 34(8):1013–1027, 8 2011.

D. Nickovic and O. Maler. AMT: a property-based monitoring tool for analog systems. In J.F. Raskin and P.S. Thiagarajan, editors, *Proceedings of the 5th International Conference on Formal modeling and analysis of timed systems (FORMATS 2007)*, volume 4763 of *Lecture Notes in Computer Science*, pages 304–319. Springer-Verlag, 2007. ISBN 3-540-75453-9, 978-3-540-75453-4.

D. Nickovic and N. Piterman. From MTL to deterministic timed automata. In *Proceedings of the 8th International Conference on Formal Modelling and Analysis of*

Timed Systems (FORMATS 2010), volume 6246 of *Lecture Notes in Computer Science*, pages 152–167. Springer, 2010.

S. Nie, L. Liu, and Y. Du. Free-fall: Hacking tesla from wireless to can bus. *Briefing, Black Hat USA*, 25:1–16, 2017.

Laurie P. and T.Z. Aziz. Security of implantable medical devices with wireless connections: The dangers of cyber-attacks. *Expert Review of Medical Devices*, 15(6):403–406, 2018. doi/10.1080/17434440.2018.1483235. PMID: 29860880.

F. Pasqualetti, F. Dörfler, and F. Bullo. Attack detection and identification in cyber-physical systems. *IEEE Transactions on Automatic Control*, 58(11):2715–2729, 2013.

H. Pearce, S. Pinisetty, P.S. Roop, M.M.Y. Kuo, and A. Ukil. Smart i/o modules for mitigating cyber-physical attacks on industrial control systems. *IEEE Transactions on Industrial Informatics*, 16(7):4659–4669, 2019a.

H.A. Pearce, Matthew M.Y.K., P.S. Roop, and S. Pinisetty. Securing implantable medical devices with runtime enforcement hardware. In *Proceedings of the 17th ACM-IEEE International Conference on Formal Methods and Models for System Design, MEMOCODE 2019*, pages 3:1–3:9. ACM, 2019b. doi/10.1145/3359986.3361200.

S. Pinisetty. *Runtime enforcement of timed properties. (Enforcement à l'éxécution de propriétés temporisées)*. PhD thesis, University of Rennes 1, France, 2015.

S. Pinisetty, Y. Falcone, T. Jéron, H. Marchand, A. Rollet, and O.L.N. Timo. Runtime enforcement of timed properties. In S. Qadeer and S. Tasiran, editors, *Proceedings of the Third International Conference on Runtime Verification (RV 2012)*, volume 7687 of *Lecture Notes in Computer Science*, pages 229–244. Springer, 2012.

S. Pinisetty, Y. Falcone, T. Jéron, and H. Marchand. Runtime enforcement of regular timed properties. In Y. Cho, S.Y. Shin, S.W. Kim, C.C. Hung, and J. Hong, editors, *Symposium on Applied Computing, SAC 2014*, pages 1279–1286. ACM, 2014a. ISBN 978-1-4503-2469-4. doi/10.1145/2554850.2554967.

S. Pinisetty, Y. Falcone, T. Jéron, and H. Marchand. Runtime enforcement of parametric timed properties with practical applications. In J.J. Lesage, J.M. Faure, J.E.R. Cury, and B. Lennartson, editors, *12th International Workshop on Discrete Event Systems, WODES 2014*, pages 420–427. International Federation of Automatic Control, 2014b. ISBN 978-3-902823-61-8. doi/10.3182/20140514-3-FR-4046.00041.

S. Pinisetty, Y. Falcone, T. Jéron, H. Marchand, A. Rollet, and O. Nguena-Timo. Runtime enforcement of timed properties revisited. *Formal Methods in System Design*, 45(3):381–422, 2014c. doi/10.1007/s10703-014-0215-y.

S. Pinisetty, T. Jéron, S. Tripakis, Y. Falcone, H. Marchand, and V. Preoteasa. Predictive runtime verification of timed properties. *Journal of Systems and Software*, 132:353–365, 2017a. doi/10.1016/j.jss.2017.06.060.

S. Pinisetty, T. Jéron, S. Tripakis, Y. Falcone, H. Marchand, and V. Preoteasa. Predictive runtime verification of timed properties. *Journal of Systems and Software*, 132:353–365, 2017b. doi/10.1016/j.jss.2017.06.060.

S. Pinisetty, P.S. Roop, S. Smyth, N. Allen, S. Tripakis, and R.V. Hanxleden. Runtime enforcement of cyber-physical systems. *ACM Transactions on Embedded Computing Systems*, 16(5s):178:1–178:25, September 2017c. ISSN 1539-9087. doi/10.1145/3126500.

S. Pinisetty, P.S. Roop, V. Sawant, and G. Schneider. Security of pacemakers using runtime verification. In *16th ACM/IEEE International Conference on Formal Methods and Models for System Design, MEMOCODE 2018*, pages 51–61. IEEE, 2018. doi/10.1109/MEMCOD.2018.8556922.

M. Renard, Y. Falcone, A. Rollet, S. Pinisetty, T. Jéron, and H. Marchand. Enforcement of (timed) properties with uncontrollable events. In M. Leucker, C. Rueda, and F.D. Valencia, editors, *Theoretical Aspects of Computing - ICTAC 2015 - 12th International Colloquium*, volume 9399 of *LNCS*, pages 542–560. Springer, 2015. ISBN 978-3-319-25149-3. doi/10.1007/978-3-319-25150-9_31.

M. Renard, A. Rollet, and Y. Falcone. Runtime enforcement using büchi games. In H. Erdogmus and K. Havelund, editors, *Proceedings of the 24th ACM SIGSOFT International SPIN Symposium on Model Checking of Software*, pages 70–79. ACM, 2017. ISBN 978-1-4503-5077-8. doi/10.1145/3092282.3092296.

M. Renard, Y. Falcone, A. Rollet, T. Jéron, and H. Marchand. Optimal enforcement of (timed) properties with uncontrollable events. *Mathematical Structures in Computer Science*, 29(1):169–214, 2019. doi/10.1017/S0960129517000123.

J. Sametinger, J. Rozenblit, R. Lysecky, and P.r Ott. Security challenges for medical devices. *Communications of the ACM*, 58(4):74–82, 2015.

F.B. Schneider. Enforceable security policies. *ACM Transactions on Software Engineering and Methodology*, 3(1):30–50, February 2000. ISSN 1094-9224. doi/10.1145/353323.353382.

D. Takahashi. Insulin pump hacker says vendor Medtronic is ignoring security risk. *Venturebeat*, 08 2011. URL https://venturebeat.com/2011/08/25/insulin-pump-hacker-says-vendor-medtronic-is-ignoring-security-risk/. (accessed 12 Feb 2019).

A. Tanveer, R. Sinha, and M.M.Y. Kuo. Secure links: Secure-by-design communications in iec 61499 industrial control applications. *IEEE Transactions on Industrial Informatics*, 2020.

P. Thati and G. Rosu. Monitoring algorithms for metric temporal logic specifications. *Electronic Notes in Theoretical Computer Science*, 113:145–162, 2005. ISSN 1571-0661. doi/10.1016/j.entcs.2004.01.029.

B. Thompson, M. Leighton, M. Korytkowski, and C.B. Cook. An overview of safety issues on use of insulin pumps and continuous glucose monitoring systems in the hospital. *Current Diabetes Reports*, 18(10):81, 2018.

C. Weaver. Patients put at risk by computer viruses. *The Wall Street Journal*, 2013.

J. Weiss. Industrial control system (ics) cyber security for water and wastewater systems. In *Securing Water and Wastewater Systems*, pages 87–105. Springer, 2014.

J. Woodcock, P.G. Larsen, J. Bicarregui, and J. Fitzgerald. Formal methods: Practice and experience. *ACM Computing Surveys (CSUR)*, 41(4):1–36, 2009.

M. Zhang, A. Raghunathan, and N.K. Jha. Medmon: Securing medical devices through wireless monitoring and anomaly detection. *IEEE Transactions on Biomedical Circuits and Systems*, 7(6):871–881, 2013.

4

Integrating Two Deep Learning Models to Identify Gene Signatures in Head and Neck Cancer from Multi-Omics Data

Suparna Saha[1], *Sumanta Ray*[2,*], *and Sanghamitra Bandyopadhyay*[3,*]

[1] *SyMeC Data Center, Indian Statistical Institute, 700108, West Bengal, Kolkata, India*
[2] *Department of Computer Science and Engineering, Aliah University, New Town, Kolkata 700160, India*
[3] *Machine Intelligence Unit, Indian Statistical Institute, 700108, West Bengal, Kolkata, India*

4.1 Introduction

The major cause of all head and neck cancers is head and neck squamous cell carcinoma (HNSCC). According to the 2015 annual report of Global Cancer Statistics, there were 500 000 new cases worldwide, which makes it one of the most common cancers (Marur and Forastiere, 2016; Siegel et al., 2015). HNSCC is a single class of multiple anatomic subsites (i.e, larynx, oropharynx, hypopharynx, and oral cavity). Most (74%) of head and neck cancers are treated with radiotherapy (Atun et al., 2015). Regionally located head and neck cancer is reasonably well controlled (90%) (Atun et al., 2015), however, the long-term survival of patients with distant metastasis (5 year survival rates) can be very poor (50%) (Atun et al., 2015; Baxi et al., 2014; Ferlito et al., 2001). Many phenotypic features and molecular alterations can give rise to carcinogenesis (Leemans et al., 2018), which allows us to recognize the heterogeneous nature of HNSCC. Therefore, a model is needed that can identify the subclasses of HNSCC patients that are at high risk of disease progression. Deep learning (DL), an advanced subclass of machine learning, can extract information from highly dimensional and complex data (Lee et al., 2009). In this work, we have proposed a framework that uses two DL models side-by-side to systematically identify gene signatures from multi-omics head and neck cancer data. The motivation behind the use of these two DL models is to integrate two sets of heterogeneous biological data and learn the integrated representation. First, we have used an autoencoder-based strategy to integrate gene expression and methylation data. We extract the integrated representation from the bottleneck layer of the autoencoder. The extracted features represent the combined representation of the gene expression and methylation data. Next, the combined extracted features stem from the integrated data, which are applied to train another DL model called capsule network. The motivation for using the capsule network is to associate the extracted feature with the class label of the data. The hidden representation of

*Corresponding Authors: Sumanta Ray, Sanghamitra Bandyopadhyay; sanghami@isical.ac.in

Applied Smart Health Care Informatics: A Computational Intelligence Perspective, First Edition.
Edited by Sourav De, Rik Das, Siddhartha Bhattacharyya, and Ujjwal Maulik.

primary capsules represents gene signatures that are responsible for discriminating cancer samples. We have also analyzed the coupling coefficients between primary and output capsules to interpret the features captured by the capsule network. We further describe the primary capsules as 'marker-capsules,' which characterizes the internal features of integrated data for specific gene signatures. Therefore, the outcome of systematically combining two DL models shows an automated identification of gene signatures that are related to specific cancer data.

An unsupervised DL model is an autoencoder. An autoencoder consists of three major parts: an encoder, a bottleneck, and a decoder. The task of the encoder is to compress the input, the bottleneck layer stores the output (a compressed representation) of the encoder, whereas the decoder tries to retrieve the input depending upon the encoded (compressed) data. Normally, the dimension of the bottleneck layer is significantly lower than the input. Hence, the aim of the encoder is to learn the relevant information as much as possible by reducing the noise so that the decoder can reconstruct the noiseless input in a better way. The autoencoder can function to reduce the dimensionality and the input stored in the bottleneck layer can be used for data visualization and other purposes.

One of the deep neural network architectures is a capsule network (CapsNet), which was initially presented in 2017/2018 and applied to digit recognition and image classification (Sabour et al., 2017). CapsNet produced appreciable advantages over the convolutional network-based approach, overcoming the fundamental and methodical issues. CapsNet provides a state-of-the-art performance on large data sets. The achievements of CapsNet were: first, to achieve higher performance in either classifying or recognizing distorted images; second, to allow for complex analysis, while retaining a simple interpretation of their primary building blocks; and third, to achieve outstanding performance, CapsNet requires significantly less training data.

4.2 Related Work

In the last two decades, gene expression signatures are widely used in omics data analysis. There are several types of gene signatures such as prognostic, diagnostic, and predictive gene signatures in the field of bioinformatics. The term 'Prognostic' signifies the prediction of the expected development (i.e. duration, description, and function) of the course of a disease, and hence, the prognostic gene signature serves a new idea into the overall outcome of a disease, irrespective of therapeutic interference. The applications of these prognostic signatures are useful in various types of tissue-specific cancers such as hepatocellular carcinoma (Hoshida et al., 2013), leukemia (Verhaak et al., 2005), and breast cancer (Nielsen et al., 2014). The diagnostic gene signature works as a biomarker that discriminates between phenotypes and analogous therapeutic conditions based on an inception point like the mild, moderate, or severe stage of a phenotype (Nguyen et al., 2015). A predictive gene signature predicts the outcome of treatment in patients, provided a specific disease phenotype. The predictive gene signature is based on the outcome of the treatment with therapeutic intervention and does not depend on the prognosis (Baker and Kramer, 2015). Hence, the information within these types of signatures is very important.

There exist several works in the literature that extensively use gene expression-based modeling for predicting the clinical outcomes of head and neck cancer (Cosma et al., 2017; Győrffy et al., 2015). Prognostic indicators for HNSCC have been studied, and nodal metastasis has usually been found to be the most important prognostic indicator (Torrecillas et al., 2018; Camisasca et al., 2009). Despite the association of prognostically important variables, absolute risk measures for individual patients remain understudied and are not usually employed in counseling patients (Steyerberg and Vergouwe, 2014; Moons et al., 2009). The evolution of clinical prediction models is vital to enhance a clinician's ability to provide absolute risk assessments (Steyerberg and Vergouwe, 2014; Moons et al., 2009). Clinicians depend on the Tumor, Node, and Metastasis groupings to carry the significance of cancer diagnosis and prognosis (Piccirillo, 1995). With the growing availability of huge national databases and computing ability, the volume of data input has grown, which allows for novel methods of data analysis (Kourou et al., 2015). One such approach involves machine learning, a subset of artificial intelligence that assists researchers in investigating large amounts of data to identify patterns for solving problems by producing predictions (Bur et al., 2019). For example, Mohd et al. (2015) predict the initial stage of head and neck cancer with fewer features by combining naïve Bayes, multilayer perceptron, Knearest neighbors, and support vector machine techniques (Mohd et al., 2015).

In another study (Ahmad et al., 2013), the objective was to build models for medical practitioners. By applying decision tree (DT) to support vector machine (SVM) and artificial neural network (ANN), they made a comparative study to identify prognostic characteristics of breast cancer. In another work, (Rajaguru and Prabhakar, 2017) extreme learning machines (ELMs) were used as a classifier for head and neck cancer analysis, and the performance of the ELM classifier was compared with the performance of both Gaussian mixer model (GMM) and multi-layer perception (MLP). In Ziober et al. (2006), an SVM classifier was used to detect Oral Squamous Cell Carcinoma (OSCC) tumors by analyzing patient expression profiles. In Aubreville et al. (2017), a novel automated approach is introduced for OSCC diagnosis using DL and convolution neural network (CNN) methods on CLE images. In another independent study, Halicek et al. (2017) proposed a deep learning-based model to identify cancer samples including HNSCC and thyroid cancer. From these related works, we observed that most of the methods used different machine learning techniques to classify cancer samples from microarray gene expression data sets. Hence, identifying gene signatures from multiple omics profiles of head and neck cancer using DL is an important task in the head and neck cancer diagnosis.

4.3 Materials and Methods

Since gene expression and methylation data have dramatically different dimensions, we first encoded them separately using two distinct encoders. Afterward, we combined the outputs of the two encoders, which pass through another encoder to construct the compressed representation of the output at the bottleneck layer. Eventually, the decoder reconstructs the input depending on the bottleneck layer output. The overall architecture of the autoencoder is described in Figure 4.1. The primary layer depicts a convolution layer. Here we took 32 primary capsules, where each capsule consisted of eight 4×1 convolution filters. The output layer consists of two 16-dimension capsules that represent two types of samples (cancer and normal) of TCGA head and neck cancer data (Please see section 4.4.1). The L2-norm of the length of the output capsule is calculated and represents the probability of the presence of that particular class.

4.3.1 A Brief Introduction of the Capsule Network

Geoffrey E. Hinton is a Canadian intellectual psychologist and computer researcher. He is one of the makers of DL and has invented many algorithms. He discovered a couple of the issues in the standard neural networks and tried to overcome them by proposing a new neural net algorithm known as capsule networks (CapsNets).

CNN, being computationally robust, can naturally recognize feature maps from the images. The CNN algorithm made it simple for a machine to do a few picture-related tasks, whether detection or classification. In CNN, the convolution layer holds a significant capacity to distinguish the features from a picture pixel. More profound CNN layers identify straightforward highlights like edges and shading. Although the presentation of CNN is great, they still have a few

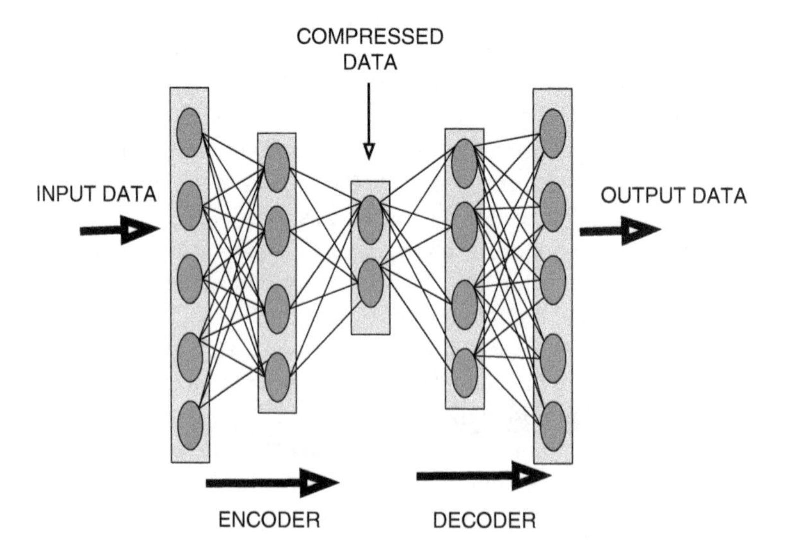

Figure 4.1 Schematic of autoencoder architecture. A representation of the physical association between the input and output layers. The compressed representation of the input data is generated at the bottleneck layer.

disadvantages. CNN requires enormous measures of information to learn. The layers in CNN decrease the spatial resolution, and the output of the networks never show any change, even with a limited amount of progress in the input sources. It can't legitimately identify with the connection of parts and requires extra components. This is where CapsNets become possibly the most important factor and beat the disadvantages of CNN.

CapsNets are the networks that can bring spatial data and more significant features to overcome the loss of data observed in pooling operations. What is the distinction between a capsule and a neuron? Capsules gives us a vector as an output with direction. For instance, if we are changing the direction of the picture, the vector will likewise move the equivalent way; the output of a neuron is a scalar amount that doesn't enlighten anything concerning the direction.

Basically, four principal segments are available in the CapsNet:

- **Matrix Multiplication:** It is applied to the input picture to covert it into vector values to comprehend the spatial part.
- **Scalar Weighting Input:** It identifies the more significant capsule that ought to get the current capsule output.
- **Dynamic Routing Algorithm:** It allows these parts to move data among one another. A more significant level capsule gets the input from the lower level. This is a recurrent process.
- **Squashing Function:** It is the last segment that condenses the data. This function converts all the data it into a vector that is near, or equivalent, to 1, and keeps the direction of the vector.

The structure comprises six layers, the initial three layers are encoders, where the function is to change the input picture into a vector; the last three layers are called decoders that are used to reproduce the picture from the data.

There are various applications of CapsNets. One of the applications of CapsNet engineering is proposed for mind tumor grouping, which takes the tumor coarse limits as additional contributions inside its pipeline to expand CapsNet's core interest. The proposed approach (Afshar et al., 2019) recognizably beats its counterparts. In Afshar et al. (2018), the use of CapsNets for brain tumor type classification was investigated. Since these networks can deal with a modest number of training samples, and units in the networks are equivalent, they solve the tumor classification problem caused by CNNs.

In study Peng et al. (2020), incorporated multi-omics data of breast cancer from the TCGA database to produce the component network of qualities, which were comprised of mRNA expression, z scores for mRNA expression, DNA methylation, and two types of CNAs. These five types of information gave extensive important information of the genes in various omics data, and the features were additionally reshaped to acquire a legitimate input for the classifier.

4.3.2 An Introduction to Autoencoders

An autoencoder is a neural network where the output layer has a similar dimensionality as the input layer. In other words, the quantity of output units in the output layer is equivalent to the number of input units in the input layer. An autoencoder recreates the information from the input to the output in an unsupervised manner and is consequently called a "replica of a neural network". The autoencoders reproduce each dimension of the input by going through the network. It might appear minor to use a neural network to repeat the input, however during the replication cycle, the size of the input is diminished into a more modest representation. The middle layers of the neural network have a smaller number of units when compared to the input or output layers. Hence, the middle layers hold a decreased representation of the input. The output is reproduced from this decreased representation of the input.

Figure 4.1 depicts the architecture of an autoencoder, which is comprised of three segments:

- **Encoder:** An encoder is a feedforward, completely connected, neural network that compacts the inputs into a latent space representation and encodes the input picture as a compacted representation in a decreased dimension. The compacted picture is the twisted adaptation of the original picture.
- **Code:** This portion of the network contains the decreased representation of the input information that is fed into the decoder.
- **Decoder:** Also a feedforward network with a structure comparative to the encoder. This network is liable for returning the input information to the first dimension from the code.

At first, the input information passes through the encoder where it is compressed and put away in the code layer; at that point, the decoder decompresses the first input from the code. The principle target of the autoencoder is to obtain an output that is indistinguishable from the input.

Note that the architecture of the decoder is the mirror representation of the encoder. This is not a prerequisite, but it is ordinarily the case. The main prerequisite is that the dimensionality of the input and output must be equivalent.

4.4 Results

4.4.1 Data Set Details

We used gene expression and methylation data of head and neck cancer obtained from TCGA (The Cancer Genome Atlas) portal. The mRNA expression and methylation 450k of head and neck cancer data sets are downloaded from TCGA (https://xenabrowser.net/datapages/). The following section includes the details of these data sets.

4.4.1.1 Gene Expression Data (Illumina Hiseq)

- Number of genes: 20 501.
- Total number of tumor samples: 530.
- Total number of normal samples: 44.

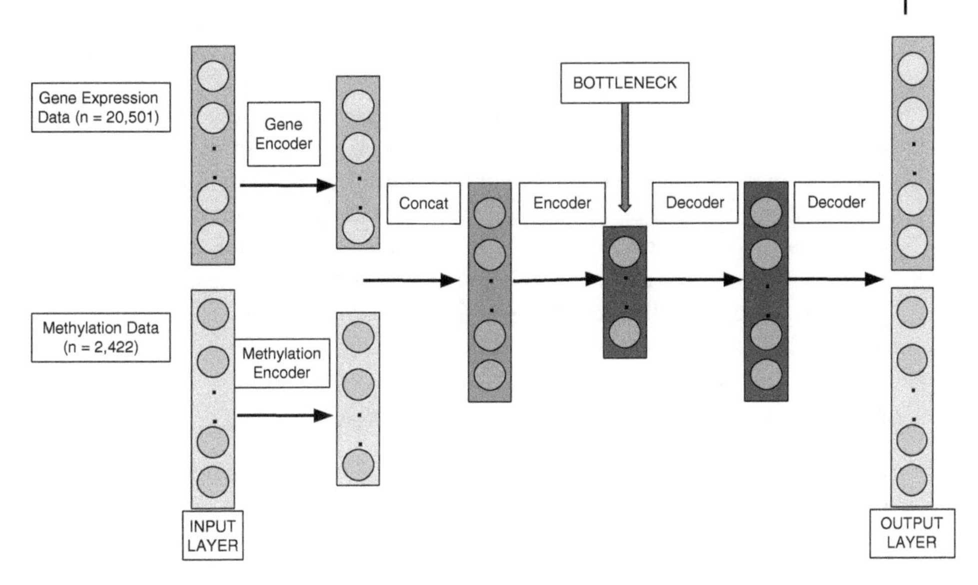

Figure 4.2 Integrative analysis of multiomics data through an autoencoder model.

4.4.1.2 Human Methylation 450K

- The total number of initial methylation-probes (i.e. CpG site probes): 485 578.
- Total number of genes with a single CpG location: 2422.
- Total number of tumor samples: 530.
- Total number of normal samples: 50.

Two categories of samples are contained in these data sets: i) treated/tumor and ii) control/normal. The number of tumor and normal samples included in both data sets are 522 and 20, respectively.

4.4.2 Architecture of Autoencoder Model

One of the most advanced unsupervised machine learning models of the artificial neural network is the autoencoder. The autoencoder is trained to regenerate the output near the initial input. Autoencoders constitute an input layer, an output layer with the same dimension as the input layer, and the hidden layer, which commonly has a dimension less than the input layer. An illustration of an autoencoder is presented in Figure 4.1 (PCA) (Wang et al., 2016; Manning-Dahan, 2018). The model operates based on the encoder-decoder model; it employs a backpropagation algorithm and places the target value similar to the input. The input is first converted into a lower-dimensional layer also known as an encoder, and then extended to reproduce the original data by what is also known as a decoder. Later, the layer is trained, and the output of the layer is sent to the subsequent layer to achieve a deeply non-linear dependencies model on the input. This method intends to decrease the dimensions of the input data. The encoded layer in the center of the autoencoder is used as a feature obtained for classification (Wang et al., 2016).

The following process is dimension reduction with an autoencoder model. In our study, we used a simple autoencoder comprised of three layers that included one hidden layer. In the beginning, we encoded gene expression data and methylation data independently using two individual encoders. Later, we connected the outputs of the two encoders, which go through another encoder to build the compact representation of the output at the bottleneck layer. Ultimately, the decoder reproduces the input depending upon the bottleneck layer output. The dimensions of the autoencoder inputs are 20 501 and 2422 for gene expression and methylation data, respectively (see Figure 4.2). In the hidden layer, we transformed the input into 64 neurons to get the best accuracy (see Figure 4.4). The output of the hidden layer displays the features obtained according to the number of neurons. The extracted data from the hidden layer has fewer dimensions or features.

4.4.3 Architecture of the Proposed Capsule Network Model

Capsule networks are an advanced model of neural networks that work on vectors rather than scalars. Considering the capsules as their defining objects, CapNets reflect the ordering of scalars (simple neurons) into vectors (the capsules). The length of the output capsule vector represents the probability of an entity's presence. CapNets consist of two layers of capsules: primary-capsules or lower-level capsules that relate to the most fundamental entities of an object and are connected to the digit-capsules or higher-level capsules, which integrate the input of primary capsules into the information that classifies the object. This mirrors the scenario that preliminary layers of neurons offer their output to enter the next layers in widespread neural networks. The detailed architecture of the proposed capsule model is shown in Figure 4.3.

In the original application, the capsule network is introduced to identify digits in MNIST (Modified National Institute of Standards and Technology) images. Each higher-level capsule (output capsule) represents the probability that one of ten digits (0 to 9) is present. The primary-capsules that produced their output as input to digit-capsules coded for adjustment, breadth, scale, depth, etc., whereas usual neurons can only code for intensity and color. Behind the improvement of classification itself, the user-friendly interpretation of primary capsules was distinguished

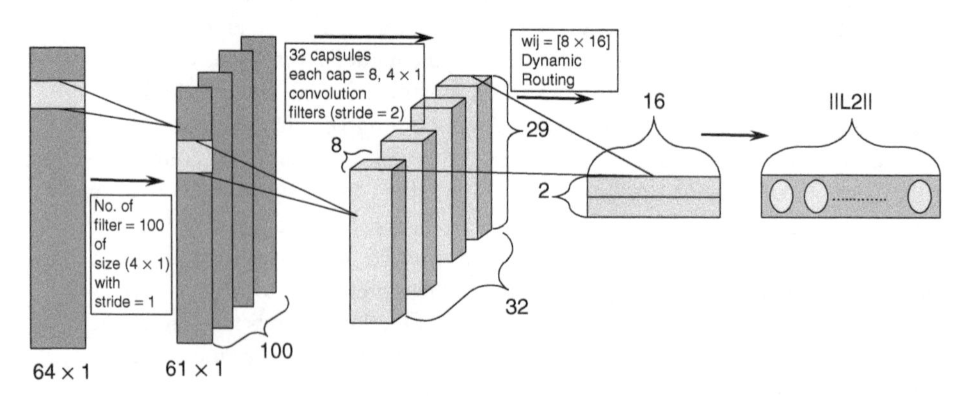

Figure 4.3 Proposed capsule network architecture. The details of the multiple layers in the proposed capsule model architecture.

as a particular benefit of CapsNets. In our study, the higher-level capsules directly correspond to a single omic. As shown in the results, we obtained primary capsules to exhibit marker genes for each omics data where such marker genes were associated. This implies that in classifying head and neck cancer samples versus the controls, primary-capsules use a human-friendly presentation. Moreover, they may infer probable sets of marker genes for each omics data set. As an essential theoretical truth, the ground invariance characteristic of CapsNets guarantees that orientation and the corresponding position of objects can be obtained under the design of CapsNets. While this property is advantageous for several reasons, a result of particular significance is that they need considerably fewer training data than other methods. The interpretation is that alternative methods often require training data to be presented with views from different angles, while CapsNets only need one datum during training. This reveals why CapsNets need considerably less training data to obtain excellent performance (Sabour et al., 2017), which are indicated in our results as well.

4.4.4 Validation of Two Deep Learning Models

We have validated the models using two steps: we first split the data into train, validation, and test sets with a ratio of 8:1:1. The validation set was used to determine the hyperparameters. We reported the accuracy value of the model in the test set. Figure 4.4 shows the training and test accuracy of the CapsNets on HNSC data. After eight epochs, the model achieves almost 99% accuracy (Figure 4.4). We applied the trained model to the test data, which also resulted in a high accuracy (test-accuracy = 0.99). This suggests that the autoencoder model perfectly extracts informative features from the data, which drives the capsule model to accurately

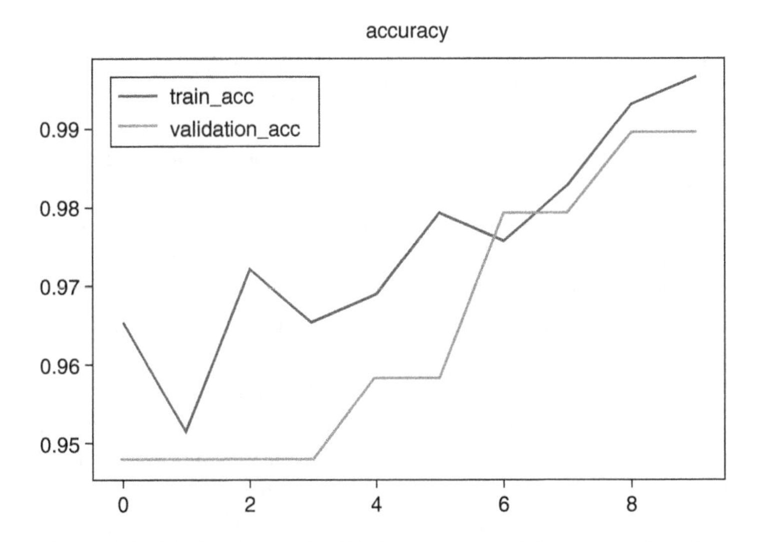

Figure 4.4 The training and validation accuracy of the proposed model.

Table 4.1 The performance (test-accuracy) of the proposed capsule and basic CNN models in small sample subsets.

Sl No.	Method	20%	40%	60%	80%	100%
1	proposed	0.924	0.968	0.969	0.985	0.987
2	CNN	0.871	0.875	0.889	0.945	0.958

learn the hidden features. Next, we compared our proposed model with a basic CNN model to evaluate the efficacy of the capsule model in learning the features compared to the CNN model. Knowing that the capsule model performed well with a small number of samples, we validated the models with different cross-validation steps. We first split the data into train, validation, and test using the same ratio as before. For training, we did not use all the training data at once but started with 5% of the training sample and kept increasing the training sample by 5% for each iteration. For each training sample, we performed a 10-fold cross-validation, in which there were 150 epochs per fold and we stored the average validation accuracy for the last 10 epochs. After each fold, the trained model was applied to the test data to get the test accuracy. Thus, we got one training accuracy (average), one validation accuracy (average), and one test accuracy value in each fold. Table 4.1 reports the results of the comparisons.

4.4.5 Gene Signatures from Primary Capsules

To determine if the primary capsules of the model have the potential to extract features with a biologically meaningful interpretation, we explored the weights of the edges that connect primary-capsules and type-capsules. These coefficients are generally known as "coupling coefficients". Therefore, for each class, we picked 100 random samples from the test set. We considered all samples if the test sample of one class had less than 100 samples. Next, we applied a trained model to each of the selected samples.

We computed the values of all the coupling coefficients for each of the 100 random samples. As the model has 32 primary-capsules that connect with two type-capsules, we obtained 2×32 coupling coefficients matrix for one test cell sample. Thus, we got an overall $100 \times 2 \times 32$ dimension matrix of coupling coefficients. We computed the average coupling coefficient across 100 samples and got 2×32 averaged coupling coefficients for each test sample. These coefficients are collected in the form of two matrices of dimension 2×32.

See Figure 4.5, for the corresponding box plots of the coupling coefficient values of the 2×32 matrices. Each box of a specific class represents a set of coupling coefficient values for all test samples. It can be seen from the figure that some of the primary capsules have larger values of coupling coefficients. For example, primary-capsule-19 and -23 have higher values for class-1 (cancer sample), while primary-capsules-9 and -17 have larger coupling coefficients in class-2 (normal

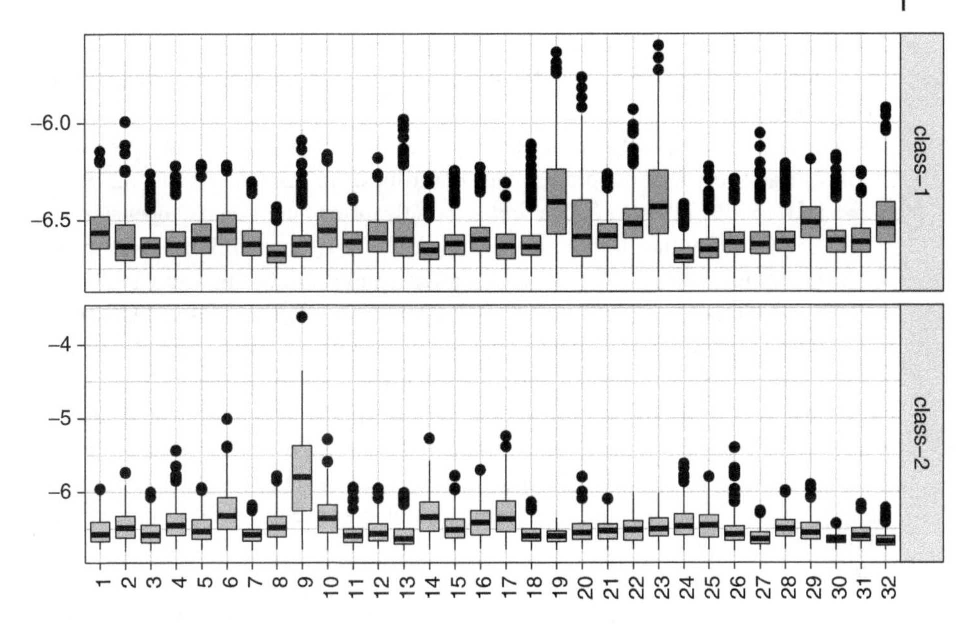

Figure 4.5 Box plots of coupling coefficient values between primary- and type-capsules.

sample). This suggests that primary-capsules-19 and 23 have prominent contributions toward extracting the features specific to cancer. A similar statement is true for primary-capsule-9 and -17 for the normal class. These primary-capsules can be treated as marker-capsules since these characterize the internal features of the integrated data for discriminating normal and cancer samples.

4.5 Discussion

Heterogeneous data that come from several simultaneous measurements consist of different biological signals. Leveraging these data resources brings up challenges in methodological developments. Developing heterogeneous data from these different experiments is crucial and requires integrating measurements from two or more experimental sources. The analysis of combined/integrated data could lead to a biologically meaningful interpretation of these complex molecular systems. Such integrated representations are most desirable in complex biological data analysis. Methods that can track how individual genes or other epigenetic effects contribute to the classification of a sample and indicate those that are beneficial, can provide an extra level of biological insight.

Here, we presented a method based on the an advanced DL architecture to address these points and provide novel and advanced solutions. Our proposed method uses the two most advanced DL models, autoencoder and CapNets, side-by-side. CapNets give full support to the automated discrimination of classes by establishing a supervised learning approach. Although the model does not establish any such approach, it supplies the greatest recent progress in DL. Although the intended primary aim is a classification with high accuracy, CapNets also allowed interpreting the input that contributed to classification. CapNets also needed training data than the methods for classifying cancer samples.

Our experiments have demonstrated that we can to classify cancer samples with a high accuracy, provided the proper CapNet settings. We have shown that our model is capable of using integrated data in the classification of unknown cancer samples. These results show that the model can outperform the conventional CNN model with respect to accuracy. The cross-validation results showed that our model can train with training samples than to a basic CNN model. In the cross-validation test, the validation accuracy of the last 10 epochs is more stable for the capsule than the CNN model. For a small number of training samples, the capsule model has a similar performance to the CNN model that was trained with a greater number of samples. The predictive performance of the capsule model trained with 40% of the samples is similar to a CNN model trained with over 75% samples. For 40% of training data, a capsule model can achieve the same accuracy as the CNN model trained with 80% of the training sample.

To understand the biological meaning of the capsules, we can associate primary-capsules with a specific class. The primary capsules are activated solely by the expression of some genes or features of samples in the data sets. We observed several primary capsules that were significantly (pval ≤ 0.001) activated for some classes repeatedly. These primary capsules may be assumed to behave as markers for the presence of those samples in the corresponding class. For thee unknown samples for which we do not know the associated class in the input data, the activated primary-capsules may indicate the predicted class in the data sets. For example, in class-1 (cancer samples), the activated primary capsules (cap-18 and cap-23) indicate the presence of cancer samples in the input data sets. Further analyses of the activated primary-capsule may identify new markers in the gene sets of the input data.

Taken together, the proposed CapNet-based model not only has a good performance on small sample training but also has the ability to explore internal lower-level features related to specific samples. Despite being applied in different fields of computer vision, the CapNet shows great potential to classify unknown cancer samples in tumor-normal, matched-pair data. It can be further explored in other fields. We believe the CapNet may be an important tool for computational biologists to explore and understand the complex patterns and features hidden within biological data sets.

Acknowledgments

This work was partially supported by the J. C. Bose Fellowship (SB/SJ/JCB-033/2016) of the Department of Science and Technology, Government of India, and by the SyMeC Project (grant number BT/Med-II/NIBMG/SyMeC/2014/Vol.II) funded by the Department of Biotechnology, Government of India.

References

P. Afshar, A. Mohammadi, and K.N. Plataniotis. Brain tumor type classification via capsule networks. In *2018 25th IEEE International Conference on Image Processing (ICIP)*, pages 3129–3133, 2018. doi: 10.1109/ICIP.2018.8451379.

P. Afshar, K.N. Plataniotis, and A. Mohammadi. Capsule networks for brain tumor classification based on mri images and coarse tumor boundaries. In *ICASSP 2019 - 2019 IEEE International Conference on Acoustics, Speech and Signal Processing (ICASSP)*, pages 1368–1372, 2019. doi: 10.1109/ICASSP.2019.8683759.

L.G. Ahmad, A.T. Eshlaghy, A. Poorebrahimi, M. Ebrahimi, A.R. Razavi, and et al. Using three machine learning techniques for predicting breast cancer recurrence. *Journal of Health & Medical Informatics*, 4:2, 2013, http://dx.doi.org/10.4172/2157-7420.1000124.

R. Atun, D.A. Jaffray, M.B. Barton, F. Bray, M. Baumann, and et al. Expanding global access to radiotherapy. *The Lancet Oncology*, 16(10):1153–1186, 2015.

M. Aubreville, C. Knipfer, N. Oetter, C. Jaremenko, E. Rodner, and et al. Automatic classification of cancerous tissue in laserendomicroscopy images of the oral cavity using deep learning. *Scientific Reports*, 7(1):1–10, 2017.

S.G. Baker and B.S. Kramer. Evaluating surrogate endpoints, prognostic markers, and predictive markers: some simple themes. *Clinical Trials*, 12(4): 299–308, 2015.

S.S. Baxi, L.C. Pinheiro, S.M. Patil, D.G. Pfister, K.C. Oeffinger, and E.B. Elkin. Causes of death in long-term survivors of head and neck cancer. *Cancer*, 120(10):1507–1513, 2014.

A.M. Bur, M. Shew, and J. New. Artificial intelligence for the otolaryngologist: a state of the art review. *Otolaryngology–Head and Neck Surgery*, 160(4): 603–611, 2019.

D.R. Camisasca, J. Honorato, V. Bernardo, L.E. da Silva, E.C. da Fonseca, P.A.S. de Faria, and et al. Expression of bcl-2 family proteins and associated clinicopathologic factors predict survival outcome in patients with oral squamous cell carcinoma. *Oral Oncology*, 45(3):225–233, 2009.

G Cosma, D Brown, M. Archer, M. Khan, and A.G. Pockley. A survey on computational intelligence approaches for predictive modeling in prostate cancer. *Expert Systems with Applications*, 70:1–19, 2017.

A. Ferlito, A.R. Shaha, C.E. Silver, A. Rinaldo, and V. Mondin. Incidence and sites of distant metastases from head and neck cancer. *Journal for Oto-Rhino-Laryngology, Head and Neck Surgery*, 63(4):202–207, 2001.

B. Győrffy, C. Hatzis, T. Sanft, E. Hofstatter, B. Aktas, and L. Pusztai. Multigene prognostic tests in breast cancer: past, present, future. *Breast Cancer Research*, 17(1):1–7, 2015.

M. Halicek, G. Lu, J.V. Little, X. Wang, M. Patel, and et al. Deep convolutional neural networks for classifying head and neck cancer using hyperspectral imaging. *Journal of Biomedical Optics*, 22(6):060503, 2017.

Y. Hoshida, A. Villanueva, A. Sangiovanni, M. Sole, C. Hur, and et al. Prognostic gene expression signature for patients with hepatitis c–related early-stage cirrhosis. *Gastroenterology*, 144(5):1024–1030, 2013.

K. Kourou, T.P. Exarchos, K.P. Exarchos, M.V. Karamouzis, and D.I. Fotiadis. Machine learning applications in cancer prognosis and prediction. *Computational and Structural Biotechnology Journal*, 13:8–17, 2015.

H. Lee, R. Grosse, R. Ranganath, and A.Y. Ng. Convolutional deep belief networks for scalable unsupervised learning of hierarchical representations. In *Proceedings of the 26th annual international conference on machine learning*, pages 609–616, 2009.

C.R. Leemans, P.J.F. Snijders, and R.H. Brakenhoff. The molecular landscape of head and neck cancer. *Nature Reviews Cancer*, 8(5):269–282. doi: 10.1038/nrc.2018.11, 2018.

T Manning-Dahan. PCA and autoencoders, 2018.

S. Marur and A.A. Forastiere. Head and neck squamous cell carcinoma: update on epidemiology, diagnosis, and treatment. In *Mayo Clinic Proceedings*, volume 91, pages 386–396. Elsevier, 2016.

F. Mohd, N.M.M. Noor, Z.A. Bakar, and Z.A. Rajion. Analysis of oral cancer prediction using features selection with machine learning. In *The 7th International Conference on Information Technology (ICIT)*, 2015.

K.G.M. Moons, D.G. Altman, Y. Vergouwe, and P. Royston. Prognosis and prognostic research: application and impact of prognostic models in clinical practice. *BMJ* 2009; 338 doi: https://doi.org/10.1136/bmj.b606.

H.G. Nguyen, C.J. Welty, and M.R. Cooperberg. Diagnostic associations of gene expression signatures in prostate cancer tissue. *Current Opinion in Urology*, 25(1):65–70, 2015.

T. Nielsen, B. Wallden, C. Schaper, S. Ferree, S. Liu, and et al. Analytical validation of the pam50-based prosigna breast cancer prognostic gene signature assay and ncounter analysis system using formalin-fixed paraffin-embedded breast tumor specimens. *BMC Cancer*, 14(1):1–14, 2014.

C. Peng, Y. Zheng, and D.S. Huang. Capsule network based modeling of multi-omics data for discovery of breast cancer-related genes. *IEEE/ACM Transactions on Computational Biology and Bioinformatics*, 17(5):1605–1612, 2020. doi: 10.1109/TCBB.2019.2909905.

J.F. Piccirillo. Purposes, problems, and proposals for progress in cancer staging. *Archives of Otolaryngology–Head and Neck Surgery*, 121(2):145–149, 1995.

H. Rajaguru and S.K. Prabhakar. Performance comparison of oral cancer classification with gaussian mixture measures and multi layer perceptron. In *The 16th International Conference on Biomedical Engineering*, pages 123–129. Springer, 2017.

S. Sabour, N. Frosst, and G.E. Hinton. Dynamic routing between capsules. In *Advances in Neural Information Processing Systems*, pages 3856–3866, 2017.

R.L. Siegel, K.D. Miller, and A. Jemal. Cancer statistics, 2015. *CA: a cancer journal for clinicians*, 65(1):5–29, doi: 10.3322/caac.21254. Epub 2015 Jan 5.

E.W. Steyerberg and Y. Vergouwe. Towards better clinical prediction models: seven steps for development and an abcd for validation. *European Heart Journal*, 35(29):1925–1931, 2014.

V. Torrecillas, H.M. Shepherd, S. Francis, L.O. Buchmann, M.M. Monroe, and et al. Adjuvant radiation for t1-2n1 oral cavity cancer survival outcomes and utilization treatment trends: Analysis of the seer database. *Oral Oncology*, 85:1–7, 2018.

R.G.W. Verhaak, C.S. Goudswaard, W. van Putten, M.A. Bijl, M.A. Sanders, and et al. Mutations in nucleophosmin (npm1) in acute myeloid leukemia (aml): association with other gene abnormalities and previously established gene expression signatures and their favorable prognostic significance. *Blood*, 106(12):3747–3754, 2005.

Y Wang, H Yao, and S Zhao. Auto-encoder based dimensionality reduction. *Neurocomputing*, 184:232–242, 2016.

A.F. Ziober, K.R. Patel, F. Alawi, P. Gimotty, R.S. Weber, and et al. Identification of a gene signature for rapid screening of oral squamous cell carcinoma. *Clinical Cancer Research*, 12(20):5960–5971, 2006.

5

A Review of Computational Learning and IoT Applications to High-Throughput Array-Based Sequencing and Medical Imaging Data in Drug Discovery and Other Health Care Systems

Soham Choudhuri[1,], Saurav Mallik[2,3], Bhaswar Ghosh[1], Tapas Si[4], Tapas Bhadra[5], Ujjwal Maulik[6], and Aimin Li[2,7]*

[1] *Center for Computational Natural Sciences and Bioinformatics, International Institute of Information Technology, 500032, Hyderabad, Gachibowli, C. N. R Rao road, India*
[2] *Center for Precision Health, School of Biomedical Informatics, University of Texas Health Science Center at Houston, TX 77030, Houston, Houston, Houston, USA*
[3] *Machine Intelligence Unit, Indian Statistical Institute, Kolkata, WB 700108, West Bengal, Kolkata, 203, BT Rd, India*
[4] *Department of Computer Science and Engineering, Bankura Unnayani Institute of Engineering, 722146, West Bengal, Bankura, Subhankar Nagar, Pohabagan, India*
[5] *Department of Computer Science and Engineering, Aliah University, 700016, West Bengal, Kolkata, Action Area II-A, 27, Newtown, India*
[6] *Department of Computer Science and Engineering, Jadavpur University, 700032, West Bengal, Kolkata, Raja Subodh Chandra Mallick Rd, India*
[7] *Shaanxi Key Laboratory for Network Computing and Security Technology, School of Computer Science and Engineering, Xi'an University of Technology, 710048, Shaanxi, Xi'an, China*

5.1 Introduction

The latest advancements in high-throughput sequencing techniques are having a great impact on biomedical challenges such as signature or biomarker detection, signature classification, single/multi-objective gene clustering, multi-omics data integration, dropout finding, predicting cancer drug response, reducing time complexity in searching specific feature/cell, drug discovery, multi-bio-molecule interaction network construction, and medical image classification that are associated with diseases such as cancer, Alzheimers, and COVID-19 (Bandyopadhyay and Mallik, 2016). Many computational tools, especially machine learning and deep learning are available to accomplish those biomedical objectives (Bandyopadhyay and Mallik, 2016). Other computing strategies such as the internet of things (IoT) are useful to health care systems in many ways. Furthermore, biomedical research provides an important portion of big data that are related to the public healthcare. There are many big biomedical data repositories (e.g. viz.ai, TCGA, ICGC, CCLE) that contain multi-omics data (Bandyopadhyay and Mallik, 2016). Analysis of high-throughput sequencing data sets (e.g. viz.ai, bulk RNA data, single cell sequencing data) is a highly active topic of research. scRNA-seq technology can

*Corresponding Author: Saurav Mallik; sauravmtech2@gmail.com

Applied Smart Health Care Informatics: A Computational Intelligence Perspective, First Edition.
Edited by Sourav De, Rik Das, Siddhartha Bhattacharyya, and Ujjwal Maulik.

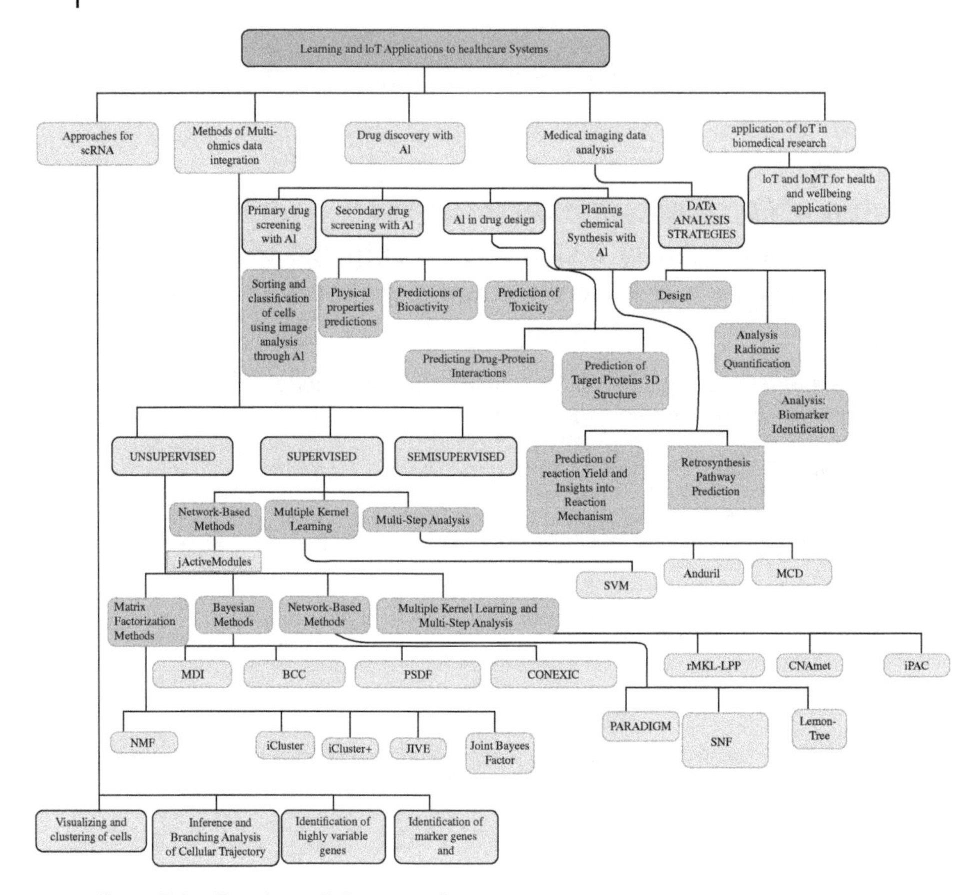

Figure 5.1 Flowchart of chapter topics.

map genotype to phenotype and is important to study cellular variations. AI helps in drug discovery. It takes around 12 years to develop a drug. With the help of computational techniques like machine learning and deep learning, it is possible to discover a drug within four years. AI helps with target selection and finding the desired chemical and physical properties of the compound. We have illustrated our chapter topics in the below flowchat (Figure 5.1).

5.2 Biological Terms

These are definitions of the biological terms that we will use throughout the chapter.

- Gene: A DNA or RNA nucleotide sequence that encodes for a protein, and it is also called a functional inheritance unit.
- DNA methylation: DNA methylation is a heritable epigenetic marker. A methyl group is covalently bound at the C-5 position of the DNA cytosine ring via DNA

Table 5.1 Methods for cell clustering and visualization.

Method	Description	Input	Reference
PCA	Creates several uncorrelated elements, reduces linear dimensionality, clarifies decreasing amounts of data variation.	Expression matrix	Pearson (1901)
t-SNE	t-distributed Stochastic Neighbour Embedding (Nonlinear dimensionality reduction).	Expression matrix	van der Maaten and Hinton (2008)
ZIFA	Uses the factor analysis technique, a method for linear dimensionality reduction that explicitly models dropout features.	Logarithm-transformed count values	Pierson and Yau (2015)
Destiny	A quick deployment of diffusion maps for R.	Expression data table	Chari et al. (2010)
SNN-cliq	Graph theory-based algorithm uses shared nearest neighbour (SNN) on a gene subset.	Expression values after normalization and log scaling (e.g. RPKM)	Xu and Su (2015)
RaceID	K-means clustering that uses an iterative method.	Raw table of gene expression.	Grün et al. (2015)
SC3	K-means clustering is first calculated, then SC3 blends several clustering results and outputs an average result (e.g. distance metrics, matrix transformation).	Expression values after normalization	Kiselev et al. (2016)
SIMLR	Learn a measure of similarity from scRNA-seq data for reduction, clustering, and visualization of dimensionality.	Estimates of raw gene expression and cell population numbers	Wang et al. (2017)

methyltransferases (DNMTs). Methylation can alter DNA segment behavior without altering the sequence.

- Gene signature: A gene signature is a single or group of genes in a cell with a particular gene expression pattern that is the result of an altered or unaltered biological process or pathogenic medical condition.
- Multi-omics: Multi-omics is a new approach that combines data sets of different omic groups during a study. Under multi-omics, the omic strategies employed are the genome, proteome, transcriptome, epigenome, and microbiome.

5.3 Single-Cell Sequencing (scRNA-seq) Data

With the launch of RNA sequencing (RNA-seq), it is now possible to measure RNA levels on genome scale to gain insights into cellular processes and illuminate many important molecular events such as alternative splicing, gene fusion, variation of single nucleotide, and differential gene expression. RNA sequencing technologies continue to advance and provide new ideas for understanding biological processes.

5.3.1 Computational Methods for Interpreting scRNA-seq Data

5.3.1.1 Visualizing and Clustering Cells

Cell classification is a significant biological issue. With the availability of genomics data, the probability of transcriptome-dependent cell similarity analysis offers an alternative cell type predictor. ZIFA (zero-inflated factor analysis) (Pierson and Yau, 2015) is a method that extends the paradigm of linear factor analysis along with conventional methods like hierarchical clustering. SNN-Cliq (Xu and Su, 2015) RaceID (Grün et al., 2015) are two important clustering methods. Others k-means clustering methods are there, e.g, single-cell consensus clustering (SC3) (Kiselev et al., 2016). SIMLR (Wang et al., 2017) is a computational method that uses multiple-kernel learning to predict similarity with particular cell populations in the gene expression matrix.

5.3.1.2 Inference and Branching Analysis of Cellular Trajectory

The trajectory analysis method is more simple than dimensionality reduction. One significant method for the pseudo time analysis of single cells is Monocle (Wang et al., 2017). It generally uses a minimum spanning tree strategy (MST) to order the cells (Magwene et al., 2003). Recently, pseudo time diffusion (dpt) was developed. This technique uses a diffusion map representation to approximate the geodesic pairwise distances between samples on the manifold of data. The trajectory is then specified as the starting distance over these distances. There are GPy and GPFlow programs for Python, and we use PSEUDOGP for R (https://github.com/kieranrcampbell/pseudogp) and DELOREAN (https://github.com/JohnReid/DeLorean).There are a few different approaches like the Ouija method (Campbell and Yau, 2016c) and the SCATER package; cells are clustered in Wishbone (Setty et al., 2016), another method by Haghverdi et al (Haghverdi et al., 2016). Methods for the bifurcation/branch identification ordering of cells are illustrated in Table 5.2.

5.3.1.3 Identifying Highly Variable Genes

Several methods identify highly variable genes (Table 5.3). A statistical approach for decomposing the total and biological variance in the generative model-based technique was suggested by Kim et al., which generally helps to classify genes with variation. BASiCS is another method that uses Bayesian techniques to model common and endogenous genes for spike-ins and the subsequent biological variability distributions (Vallejos et al., 2015).

Table 5.2 Methods for the bifurcation/branch identification ordering of cells.

Method	Description	Input	Reference
PQ-trees	Minimum spanning tree of data is ordered by samples, using a PQ-tree construction.	Expression matrix	Magwene et al. (2003)
Monocle2	In the transcriptome space, a principal graph is embedded, pseudo-time is depicted with distance along with the specified graph from a initiating cell.	Gene list, batch effect formula, expression matrix, options of dimensionality reduction (method, number of dimensions)	Qiu et al. (2017)
Wishbone	Diffusion maps (using way points) on the reduced k-NN graph.	Expression data table, initiating cell, number of nearest k-neighbours	Setty et al. (2016)
Wanderlust	Geodesic distance of the heuristic k-NN graph.	Expression data table	Bendall et al. (2014)
DPT	Based on spectral embedding, diffusion components are averaged for each sample and used as a gap between samples.	Expression data table, variance of Gaussian kernel, initiating cell	Mo et al. (2013b)
GPLVM	Assume certain smooth functions are followed by genes and deduce time as the latent parameter.	Co-variance function, expression matrix or dimensionality reduction, optional priors, optional co-variance function hyperparameters	Macaulay et al. (2016) and Campbell and Yau (2016a)
Ouija	Time is treated as a latent variable provided a limited number of sigmoidal genes around the trajectory.	Optional priors that switch time and direction, assumed switch-like genes list, expression matrix	Campbell and Yau (2016c)
Branching analysis Wishbone	Clusters detours between cells relative to a beginning cell in terms of the pseudo-time, two branches are detected.	Expression matrix	Setty et al. (2016)
Anti-correlation clustering	Branch points are defined and cells can either be segmented from the two branches or from the trunk to belong to them.	Expression data table	Mo et al. (2013b)
GPfates/OMGP	Model data as a mix of processes that are ongoing. Each cell has a subsequent probability of generating each branch.	Expression matrix	Lönnberg et al. (2017)
Monocle	The definition of branches to which cells are assigned depicted as a main graph directly attached to the expression data.	Expression data table, gene list	Lock and Dunson (2013)
Mpath	Minimum spanning tree in the landmarks' neighborhood graph.	Expression matrix	McCarthy et al. (2016)

Table 5.3 Gene-level analysis tools (identifying differentially expressed genes).

Method	Description	Input	Reference
SCDE	Compares two groups of single cells with the Bayesian technique, takes into account heterogeneity in scRNAseq findings due to bias in dropout and amplification.	Count values of gene expression data profile	Kharchenko et al. (2014);
MAST	Uses a two-part generalized linear model adjusted for the detection rate of cells.	Normalized gene expression values	Andrews and Hemberg (2016)
M3Drop	To grasp differential expressions, a Michaelis-Menten modeling of dropouts is used.	Count values of gene expression data profile	Finak et al. (2015)
scDD	A Bayesian modeling framework to identify genes within modes that are expressed differently and/or represent a different number of modes or a different proportion of cells.	Gene expression values after normalization and log-scaling	Korthauer et al. (2016)
SINCERA	Recognizes DE genes, based on basic statistical tests such as the Wilcoxon rank sum test and t-tests.	Count values of gene expression data profiles	Guo et al. (2015)
DESeq2	A Wald or LR significance test applied to a GLM for each gene, uses a shrinkage estimation for dispersion and fold modifications.	Count values of gene expression data profiles	Love et al. (2014)
EdgeR	A negative binomial distribution fit, estimates dispersion by conditional maximum likelihood for each gene, defines differential expression using an appropriate test optimized for over-dispersed results. Supports arbitrary linear models.	Count values of raw gene expression data	Robinson et al. (2009)
Brennecke et al	Biological heterogeneity of genes is inferred after quantifying technical noise centered on the square of variance coefficient (CV2) of spike-in molecules.	Count values of raw gene expression for both spike-ins and endogenous genes	[115]

Kim et al.	It provides a statistical method based on a generative model to decompose the total variance into the technological and biological variance.	Count values of raw gene expression for both spike-ins and endogenous genes	[116]
BASiCS	A Bayesian strategy that jointly models spike-ins and endogenous genes. In relation to high (or low) variable genes, posterior probabilities are given.	Count values of raw gene expression for both spike-ins and endogenous genes	Vallejos et al. (2015)
scLVM	Dissects observed heterogeneity into multiple sources using a Gaussian Process Latent variable model to remove confounding variance factors such as cell cycle variatios.	Count values of gene expression data profiles and a collection of genes related to the latent factor	Buettner et al. (2015)
Combat	Centered on an empirical Bayesian system, it eliminates known batch effects.	Counts of gene expression from batches after normalization and log scaling	Leek et al. (2012)
OEFinder	Identifies possible artifacts (ordering effects), developed using orthogonal polynomial regression by the Fluidigm C1 platform.	A number of genes (and P-values) that the artifact is influenced by	Lin et al. (2013)
RUVSeq	Adjustments performed on a set of control genes (e.g. spike-ins) or samples (viz.ai, replication libraries) to provide nuisance technological impacts through factor analysis.	Count values of gene expression data profiles and a collection of genes for regulation, spike-ins, or replicate libraries	Vallejos et al. (2015)
Monocle SwitchDE	Spline regression using VGAM.	Expression matrix, gene list	Lock and Dunson (2013)
SwitchDE	Identify genes depicted as sigmoid curves over pseudo time.	Expression matrix	Campbell and Yau (2016b)
ImpulseDE	Determine genes that follow an impulse model.	Expression data table	Sander et al. (2016)
GP Regression	Identify genes that follow any smooth non-linear function.	Expression matrix	GPy, GPFlow, others

5.3.1.4 Identifying Marker and Differentially Expressed Genes

This is an important and significant analysis technique for scRNA-seq research. DESeq2 (Love et al., 2014) and EdgeR (Robinson et al., 2009) methods are commonly used in bulk RNA-seq experiments. By adjusting and fitting a GLM for each gene, DESeq2 recognizes differentially expressed genes. M3Drop and Michaelis-Menten modeling are other techniques that use dropouts for scRNA-seq, and this method is used to classify genes that break up (Andrews and Hemberg, 2016). SCDE is a significant Bayesian method for comparing two single cell groups, taking into account the heterogeneity of the dropout and amplification bias scRNAseq results, and uses a mixture of the two components to determine the difference in terms between conditions (Kharchenko et al., 2014). Simple statistical tests such as the Wilcoxon test and t-test (Guo et al., 2015) identify differently expressed genes. A newer method, scDD, determines the genes that have modified the overall distribution of values between conditions relative to the method.

5.4 Methods of Multi-Omic Data Integration

A new era is arriving, which proposes a model of health care and treatment for individuals, with objectives tailored to each individual patient. Under this scheme, a patient's clinical profile and molecular profiles are individually able to push for diagnosis. There is a tremendous effort to obtain genomics data in bio-specimen-based multi-platforms. We are first going to talk about data integration calculations without supervision focusing on the lattice factorization strategy, Bayesian strategies, and strategies-based system. We are also going to survey the strategy of the mentioned data integration in depth, including with network-based techniques. We have illustrated different tools or methods for multi-omics data analysis in Figure 5.2.

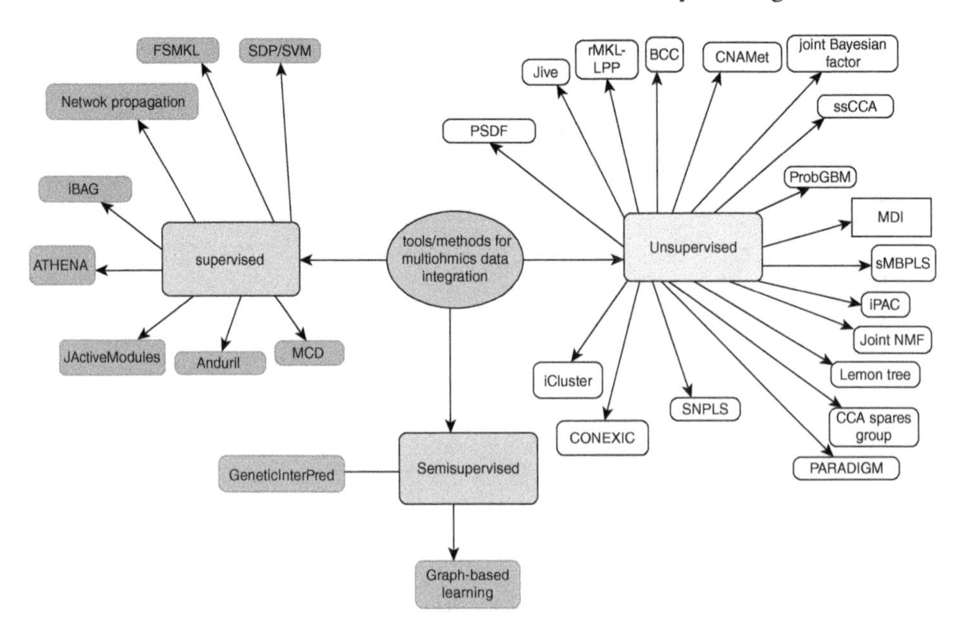

Figure 5.2 Tools or methods for multi-omics data analysis.

5.4.1 Unsupervised Data Integration Methods

This approach is called unsupervised data integration, which is inferred from input data sets without labeled output variables. The distinct approaches are detailed in Table 5.4.

5.4.1.1 Matrix Factorization Methods

Matrix factorization is a class of algorithms for collaborative filtering that are used in recommendation systems. Matrix factorization algorithms operate by decomposing the matrix of user-item interactions into two rectangular matrices of lower dimensionality.

Joint Non-Negative Matrix Factorization (NMF) A simple and significant tool for data integration that concentrates on projecting the variation between the data-dimensional space to be reduced. Zhang et al. (2011) has given an NMF method for multi-omics data integration. This NMF depends upon the decomposition of a non-negative matrix into non-negative factors and a non-negative load:

$$min||X - WH||^2, \ W \geq 0, \ H \geq 0. \tag{5.1}$$

Here, **(X)** is the matrix of an mRNA transcriptome or other omics. The data set has $M \times N$ dimensions. For the $M \times$ K-dimensional matrix, W is a common factor, and the coefficient matrix **(H)** has dimension $K \times N$.

iCluster iCluster (Shen et al., 2012) assumes regularized latent variables and is like W in NMF (Equation 5.1) but without the constraints of non-negative. H refers to a loading factor (coefficient), the sparsity paid for separate data types with different types of punishment functions. To represent the term error, iCluster uses E and the decomposition of the underlying equations is:

$$X = WH + E. \tag{5.2}$$

iCluster+ The upgraded iCluster+ extends iCluster by hypothesizing modelling methods inside of different data platforms for the **(X)** and W relationships (Equation 5.2). It allows for various types of data, including continuous, binary, categorical and sequential data with various models. Unlike NMF data, we don't need a non-negative input data for both iCluster and iCluster+.

Other important methods are Joint and Individual Variation Explained (JIVE) (Lock and Dunson, 2013) and Joint Bayes Factor.

5.4.1.2 Bayesian Methods

Bayesian Methods are important methods for medical data computation.

Multiple Data Set Integration (MDI) It enables use of the Dirichlet-Multinomial Allocation (DMA) mixture model (Kirk et al., 2012). In addition to bi-clustering across multiple data sets, MDI generally obtains a single-dimensional cluster across all input data sets after learning the similarity of clusters across the data sets. Prob_GBM is a probabilistic model for constructing a network of patient similarities where patients are represented by a node and the edge between patients (Cho and Przytycka, 2013) corresponds to the phenotypic similarity.

Table 5.4 Unsupervised data integration methods.

Method	Data type	Output	Stat method	FS method	Reference
Joint NMF	Multi-data	Subset of genes(modules).	Matrix factorization	NA	Zhang et al. (2011)
iCluster	Expression data (EXP), Copy number data (CNV)	Cluster	matrix factorization	L1 penalty	Shen et al. (2012)
iCluster+	Multi-data	Cluster	Matrix factorization	L1 penalty	Mo et al. (2013a)
JIVE	Multi-data	Shared and unique factors	Matrix factorization	L1 penalty	Lock and Dunson (2013)
Joint Bayes Factor	EXP, Methylation data (MET), Copy number data	Shared and unique factors	Matrix factorization	Prior student-t sparseness promoting	Ray et al. (2014)
ssCCA	Sequence data	Operational taxonomic units and clusters	Canonical Correlation Analysis	L1 penalty	Chen et al. (2012)
CCA sparse group	Two types of data	Group features with weights	Canonical Correlation Analysis	L1 penalty	Lin et al. (2013)
sMBPLS	Multi-data	Group features as modules	Partial Least Squares	L1 penalty	Li et al. (2012)
SNPLS	EXP, drug response, gene network info	Gene-drug co-module	Partial Least Squares	Network-based penalty	Chen and Zhang (2016)
MDI	Multi-data	Cluster	Bayesian	NA	Kirk et al. (2012)
Prob_GBM	Expression data, copy number data, miRNA, SNP	Cluster	Bayesian	NA	Cho and Przytycka (2013)
PSDF	EXP, copy number data	Cluster	Bayesian	Binary indicator– > likelihood of feature	Yuan et al. (2011)

BCC	Expression data, methylation data, miRNA, proteomics	Cluster	Bayesian	NA	Lock and Dunson (2013)
CONEXIC	Expression data, CNV	Group genes associated with modulators	Bayesian	NA	Akavia et al. (2010)
PARADIGM	Multi-data	Gene score and significance in each pathway	Pathway networks	NA	Vaske et al. (2010)
SNF	Expression data, methylation data, miRNA	Cluster	Similarity network fusion	NA	Wang et al. (2014)
Lemon-Tree	Expression data, one of miRNA/CNV/methyl	Association network graphics	Module network	NA	Bonnet et al. (2015)
rMKL-LPP	Multi-data	Cluster	Multiple kernel learning	Dimension reduction metric, Locality Preserving Projections (LPP)	Speicher and Pfeifer (2015)
CNAmet	Expression data, methylation data, copy number data	Scores and p-values of genes	Multi-step study	NA	Louhimo and Hautaniemi (2011)
iPAC	Expression data, copy number data	Subset of genes	Multi-step study	Several filtering steps including common aberrant genes, in-cis correlation and in-trans functionality	Aure et al. (2013)

Other Methods There are some Bayesian methods that are important, these are (i) Patient-Specific Data Fusion (PSDF), which is dependent upon the Dirichlet Process model's two-level hierarchy (Yuan et al., 2011); (ii) Bayesian Consensus Clustering (BCC), a clustering technique that models the dependency and heterogeneity of multiple data sources at the same time (Lock and Dunson, 2013); and (iii) COpy Number and EXpression In Cancer (CONEXIC), which is a network-based Bayesian technique for combining CNV and gene expression data (Akavia et al., 2010).

5.4.1.3 Network-Based Methods

This method is used to identify the module, a symbolic representation of disease-related techniques (Vaske et al., 2010; Wang et al., 2014; Bonnet et al., 2015).

Pathway Representation and Analysis by Direct Reference on Graphical Models (PARADIGM) This is a graphical model that uses a probabilistic framework for adding a specific patient's genetic variation to the conclusion. It incorporates a curated pathway interaction between genes (Vaske et al., 2010).

Similarity Network Fusion (SNF) This technique seeks to discover clusters of subgroups of patients. By building a network of samples (rather than genomics features) for each data type, SNF integrates multiple data types and fuses the resulting networks into a single comprehensive network (Wang et al., 2014).

Lemon-Tree This is an unsupervised approach that focuses on reconstruction of the (Bonnet et al., 2015) network module. Lemon-Tree helps classify consensus modules and upstream regulatory programs using ensemble methods after identifying co-expressed clusters from the expression data matrix.

5.4.1.4 Multi-Step Analysis and Multiple Kernel Learning

Usually, multi-step (or multi-stage) methods are used to find similarities between data types and then between the different features or phenotypes.

Regularized Multiple Kernel Learning Locality Preserving Projections (rMKL-LPP) With this approach, it is possible to explore multiple integration of omics data such as gene expression, DNA methylation, and micro-RNA expression profiles (Speicher and Pfeifer, 2015). This approach is an extension of Multiple Kernel Learning Dimensional Reduction (MKL-DR).

CNAmet This is a multi-step integration platform for data on gene expression, DNA methylation and CNV (Louhimo and Hautaniemi, 2011). CNAmet's essential aim is to identify characteristics that are either both amplified and up-regulated, or both erased and down-regulated.

In-Trans Process Associated and Cis-Correlated (iPAC) This is a multi-step process by which in-cis-related genes can be defined by combining CNV data, gene expression, and genes associated with in-trans biological processes (Aure et al., 2013).

5.4.2 Supervised Data Integration

To test the models, supervised methods (Table 5.5) consider the phenotype labels of samples (disease or normal) and apply machine training approaches. Methods of supervised data integration are constructed by the knowledge of available known labels from training omics data.

5.4.2.1 Network-Based Methods

jActiveModules This is a network-based approach that integrates data from protein-protein interaction, gene expression, and protein-DNA interaction into the hot-spots (Ideker et al., 2002) underlying network. JActiveModules identifies the highest-scoring sub-network in a complete network of molecular interactions via a random sampling method and iterative simulations, which results in a more biologically interesting performance.

5.4.2.2 Multiple Kernel Learning

Support Vector Machine (SVM) This is a kernel-based data integration framework, Lanckriet et al. (2004). A particular kernel function that determines similarities between pairs of entities defines each data set. The kernel functions are then directly combined using SDP (Semi-definite Programming) techniques, derived from multiple omics data, to reduce the integration problem to a convex optimization problem.

5.4.2.3 Multi-Step Analysis

Essentially, this multi-step analysis method is recognized as a two-stage model. The first-stage model is a regression model designed to divide the data into smaller segments for gene expression, including the component's critical methylation, the main CNV component. From the previous two, no other components are identified.

Multiple Concerted Disruption (MCD) It allows the incorporation of the status of DNA methylation, CNV and allelic (loss of heterozygosity) to identify genes that serve primary pathway nodes as well as genes with prognostic significance (Chari et al., 2010). As a pioneering work in the field of data integration, by integrating parallel analysis into genomic and epigenomic layers, MCD offers a biologically acceptable way of wisely choosing genes. It is a filtering stage to finalize a community of genes.

Anduril This is a tool in bioinformatics designed to deliver integrative results in a study for biologists to better understand data from different platforms (Ovaska et al., 2010). It is a versatile analytical tool that helps integrate multiple data formats to determine genes and loci with a high survival effect.

5.4.3 Semi-Supervised Data Integration

The semi-supervised integration method is situated between supervised and unsupervised methods (Table 5.6). It takes two samples, labeled and unlabeled, to develop learning algorithms.

Table 5.5 Supervised data integration tools.

Method	Data type	Output	Stat method	FS method	Reference
ATHENA	EXP, CNV, MET, miRNA	Final model with patient index	Grammatical Evolution Neural Networks(GENN)	Neural Networks	(Kim et al., 2013b)
jActiveModules	EXP, PPI, protein-DNA interactions	Sub-network (network hotspots)	Network simulated annealing	NA	(Ideker et al., 2002)
Network propagation	Gene expression, mutation, PPI	Propagated network relative to differential expression of genes	Network	NA	(Ruffalo et al., 2015)
SDP/SVM	EXP, protein sequence, protein interactions, hydropathy profile	Linear classifier depending upon on the combination of kernels	SDP/SVM	Recommends CCA (canonical correlation analysis)	(Lanckriet et al., 2004)
FSMKL	EXP, CNV, Clinic features (ER status)	Linear classifier depending upon on the combination of kernels.	Multiple kernel learning	SimpleMKL (gradient descent method)	(Seoane et al., 2013)
iBAG	Multi-data	Subset of genes	Multi-step analysis	Bayesian	(Jennings et al., 2013)
MCD	MET, CNV, LoH	Subset of genes	Multi-step analysis	NA	(Chari et al., 2010)
Anduril	EXP, MET, miRNA, exon, aCGH, SNP	Comprehensive report	Multi-step analysis	NA	(Ovaska et al., 2010)

Table 5.6 Semi-supervised data integration tools.

Method	Data type	Output	Stat method	FS method	Reference
GeneticInterPred	EXP, PPI, protein complex data	Genetic interaction labels	Graph integration	NA	You et al. (2010)
Graph-based learning	EXP, CNV, MET, miRNA	Patient scores for classification purpose	Graph integration	NA	Kim et al. (2013a)

5.4.3.1 GeneticInterPred

This is a method for predicting genetic interactions by integrating data interactions between protein-protein, protein complexes, and gene expression (You et al., 2010). Multi-level genomics data sources (including CNV, gene expression, methylation, and miRNA expression) using graph-based SSL (Kim et al., 2013a) are used for the molecular classification of clinical outcomes in another pilot process.

5.5 AI Drug Discovery

Drug discovery is a very lengthy process comprising of four major steps. (i) Selection of proper target and validation; (ii) compound screening; (iii) preclinical studies; (iv) clinical trials. Many pharma-companies are involved in discovering new drugs, but 90% of all drug discovery programs are failing. Due to huge amount of data available in medical and health care industry, AI plays a big role in data integration and interpretation. AI helps to determine drug targets, find expected molecules from data libraries, provide suggestion for chemical and physical modifications, etc (Table 5.7). The first drug (named DSP-1181) designed by Professor Andrew Hopkins of Exscientia using AI is in Phase I trials. This drug was developed for the treatment of obsessive-compulsive disorder during a joint venture between Exscientia and Sumitomo Dainippon Pharma. Usually, it takes about five years for drug production to get to trials, but AI only required one year for the purpose. From this example, and numerous other new initiatives, we can say that AI has huge potential to be implemented successfully in health sectors.

5.5.1 AI Primary Drug Screening

5.5.1.1 Cell Sorting and Classification with Image Analysis

AI is good at image recognition. For the classification of target cells we can use AI-based technology. We train AI models quickly, and then the model can recognize the different features of cell types. As an example, the classification of cancer cells. Least square support vector machine (LS-SVM) shows the highest accuracy

Table 5.7 List of AI-based computational tools for drug discovery.

Tools	Description	Reference
AlphaFold	Protein 3D structure prediction.	
Chemputer	A more structured format for documenting chemical synthesis procedures.	Steiner et al. (2018)
DeepChem	A Python-based AI platform for predictions of drug discovery tasks.	Ramsundar et al. (2019)
DeepNeuralNet-QSAR	Molecular activity predictions.	Xu et al. (2017)
DeepTox	Toxicity predictions.	Mayr et al. (2016)
DeltaVina	A scoring feature for protein-ligand binding affinity re-scoring.	Wang et al. (2017)
Hit Dexter	ML models for molecule prediction that could react to biochemical assays.	Stork et al. (2019)
Neural Graph Fingerprints	Prediction of properties for novel molecules.	Duvenaud et al. (2015)
NNScore	For protein-ligand interactions, a neural network-based scoring mechanism.	Durrant and McCammon (2011)
ODDT	A robust chemoinformatics and molecular modelling toolkit.	Wójcikowski et al. (2015)
ORGANIC	A powerful method for molecular generation to build molecules with desired characteristics.	Sanchez-Lengeling et al. (2017)
PotentialNet	Ligand-binding prediction of affinity on the basis of a convolution Neural Network (CNN).	Feinberg et al. (2018)
PPB2	Poly-pharmacology prediction.	Awale and Reymond (2018)
QML	A Python toolkit for quantum ML.	
REINVENT	Utilizing RNN (recurrent neural network) and RL (reinforcement learning), molecular de novo architecture.	Olivecrona et al. (2017)
SCScore	A scoring feature for the assessment of a molecule's synthesis complexity.	Coley et al. (2018)
SIEVE-Score	An improved method of structure-based virtual screening via interaction-energy-based learning.	Yasuo and Sekijima (2019)

for regression and classification techniques. Image activated cell sorting (IACS) is used to calculate the mechanical, optical and electrical properties of cells for flexible cell sorting. Recently, AI was used to interpret computerized electrocardiography (ECG). Available digital ECG data and deep learning (DL) can significantly enhance the accuracy and scalability of automated ECG analysis.

5.5.2 AI Secondary Drug Screening

5.5.2.1 Physical Properties Predictions

A significant goal in drug design is selecting a drug candidate that displays multiple desirable properties, particularly bio-availability, bio-activity, and appropriate toxicity. Taking these characteristics into account, the molecular representations used in an computational drug design algorithm include a molecular fingerprint, a simplified molecular input line-entry method (SMILES) series, possible energy measurements (e.g. from ab initio calculations), varying weight molecular graphs, Coulomb matrices, molecular fragments or bonds, and 3D atomic co-ordinates. These inputs are generally used to train DNN. We use DNN in various stages, namely the generative and predictive stages. Reinforcement learning (RL) is a very effective method to predict molecules. We use SMILES in the generative stage of DNN as the input and train for chemically decent SMILES strings; the properties of the molecules are trained in the predictive stage. Initially, we train two stages separately with a supervised learning algorithm; the bias can be used to predict the result when the two phases are trained together with rewards or punishment for certain properties.

5.5.2.2 Predictions of Bio-Activity

Small local changes to a drug candidate and its effect on the molecular properties and bio-activity of the molecule are investigated in the Matched Molecular Pair (MMP) analysis technique. This method is generally used to study quantitative structure-activity relationship (QSAR). Three machine learning methods, random forest, gradient increases engine, and DNNs are used to do a new transformation into fragments. It is noted that in predicting compound activity, DNN has a better overall performance than RF or GBM. There are a few methods to predict the bio-activity of the drug candidate. As an example, Tristan et al. predicted the 'drug target site' signature using a network coding convolutional graph with discrete chemicals into sustainable space latent vectors (LVS). LVS predicts the model's differential affinity and other binding properties.

5.5.2.3 Prediction of Toxicity

A significant parameter in drug production is the toxicological profile of compounds. In the preclinical phase, toxicity optimization is a costly and timely task (Blomme and Will, 2015; Deshmukh et al., 2012). DeepTox (ML algorithm) (Mayr et al., 2016) (Table 5.1), is delivering good results in Tox21 Data Challenges (Krewski et al., 2010). In a typical test case, the DeepTox algorithm showed excellent accuracy in predicting compounds' toxicology (Mayr et al., 2016).

5.5.3 AI in Drug Design

5.5.3.1 Prediction of Target Protein 3D Structures

Predicting 3D structures of target proteins is the most important thing for structure-based drug discovery (Chan et al., 2019, 2018). Now, many researchers are designing new drug molecules according to the 3D chemical environment and structure of a target protein's ligand-binding site. The AI tool AlphaFold is used to anticipate the 3D structure of a protein drug target and performed superbly well in a recent Crucial Assessment of Protein Structure Prediction contest. AlphaFold correctly predicts 25 out of 43 structures using only the primary protein sequence. To find structures in line with the prediction, AlphaFold investigates the protein structure landscape.

5.5.3.2 Predicting Drug-Protein Interactions

A hybrid method of molecular mechanics (MM) is useful in drug discovery for predicting protein-ligand (drug) interactions. These methods take into account the quantum effects at the atomic level of device simulation, thereby providing better precision than classical MM methods. The application of QM calculations to AI methods requires trade-offs between QM precision and MM's time-cost model.

5.5.4 Planning Chemical Synthesis with AI

5.5.4.1 Retro-Synthesis Pathway Prediction

Retro-synthesis is a design method used for organic synthesis. AI has given us the facility (Coley et al., 2018) to find an appropriate chemical synthesis pathway for the drug candidate once it has been digitally screened for its possible bio-activity and toxicology profile. It is now possible to use the trained ML method on empirical data (i) to predict the transformation probability in a certain branching position and (ii) help with the selection of random steps. For a molecules retro-synthesis that has been submitted for analysis, this node is operated iteratively to find the transformation with the highest score and will eventually identify a possible precursor to full reaction pathways.

5.5.4.2 Reaction Yield Predictions and Reaction Mechanism Insights

ML algorithms are very effective for predicting the outcome of a reaction on the basis of the reactant molecule's physical and chemical properties. Predicting the outcome of a complex chemical reaction has been a major challenge. Quantum chemistry techniques can theoretically solve this problem such as the Hartree-Fock method, density functional theory, and semi-empirical approaches. Several studies have recently been published in this field using AI algorithms to automate, boost, and generalize yield prediction. Doyle and Dreher say that ML can be used to predict the outcome of the BuchwaldHartwig coupling reaction.

5.6 Medical Imaging Data Analysis

Disease diagnostic imaging such as Magnetic Resonance Imaging (MRI), X-Ray, Computed Tomography (CT), and Ultrasound plays an important role in health care. Nowadays, ML and AI techniques are widely used in clinical practice (Allison et al., 2006). AI/ML techniques help to detect and classify different kinds of disease. Segmentation is an important image analysis task to detect abnormalities in organs and body parts. Intelligent search techniques like Particle Swarm Optimization

(PSO) and Glowworm Swarm Optimization (GSO) have been successfully applied to the segmentation of medical images (Aerts, 2017). Recently, applications of DL in image classification, segmentation, registration, and object detection have gained in popularity because of higher success rates than traditional ML techniques.

Brain MRI segmentation is an significant medical image analysis technique for brain lesion detection. Multi-level thresholding is a segmentation method in which suitable threshold values are selected to separate the tumors from healthy tissues in the brain. Allison et al. (2006) developed a segmentation method in which Grammatical Swarm was used to search the threshold values through entropy maximization for lesion and tumor detection. Unsupervised methods such as clustering techniques are widely used in brain MRI segmentation. Aerts (2017) developed a hard-clustering technique with generalized opposition-based GSO for lesion segmentation in brain MRI. T. Si et al. developed artificial neural network (ANN)-based supervised segmentation method using extracted statistical features from brain MRIs. In this study, a classifier was developed using Grammatical Bee Colony to segment the brain lesions in MRI; the stationary wavelet features of MR images are used as inputs to the classifier. Breast cancer causes a huge number of the deaths from cancers in women. Dynamic contrast-enhanced (DCE)-MRI is used for detecting lesions and tumors in the breast and the characterization of lesions as benign or malignant. In studies, the statistical learning model Markov Random Field (MRF) is generally used to segment the breast lesions in DCE-MRI. K.S. Sim et al. developed a computer-aided detection auto-probing (CADAP) system to detect breast lesions using a spatial-dependent discrete Fourier transformation. A. Gubern-Mérida et al. used an expectation-maximization (EM) method for fibroglandular tissue segmentation in the breast area. H. Jiao et al. applied a deep convolution neural network for automatic breast segmentation and mass detection in DCE-MRI. Wang et al. suggested a way to identify the breast into fatty, glandular, tumor and muscle in DCE-MRI using a SVM. Miotto et al. (2017) proposed a hybrid intelligent system using a type-II fuzzy set, pulse coupled neural network (PCNN), and SVM to classify breast MRIs into cancer or not cancer. Zhou. et al. (2017) proposed a tumor classification method in breast DCE-MRI using textural, wavelet features and a committee of SVM. A mask-guided hierarchical system was developed using CNN for breast tumor segmentation in DCE-MRI by Wang (2016).

5.6.1 Analysis: Radio-Mic Quantification

In radiology, AI can perform a comprehensive quantification of a network's characteristics. This method can convert 3D radiology image descriptors to high-dimensional phenotypes. This approach, called radio-mics, uses engineered features and/or in-depth learning.

5.6.2 Analysis: Bio-Marker Identification

To classify highly predictive/prognostic bio-markers, quantitative analysis of medical image data requires the mining of vast numbers of imaging features. Biomarker identification and analysis is important for diagnosis of complex diseases.

5.7 Applying IoT (Internet of Things) to Biomedical Research

5.7.1 IoT and IoMT Applications for Healthcare and Well-Being

5.7.1.1 Wireless Medical Devices

Wireless and sensor networks and devices are rapidly increasing. These types of networks and devices provide automated, continuous, and real-time physiological signal measurement and performs unrestricted data processing and functioning as fundamental components of an intelligent health solution. Vital elements such as blood pressure, skin conductance, heart rate variability, body temperature, breathing rate, blood sugar, oxygen saturation, and related activities can be monitored and measured using properly chosen sensors that can be placed on clothing or directly on the body. Various sensor technologies can be used to calculate human physiology at any given point (Jeong et al., 2019). It is easy to measure the pulse rate, which has become a routine measure, from the PPG signal (photoplethysmography) (Mendelson et al., 2013). IoT is playing a big role in the healthcare industry in different ways.

5.8 Conclusions

Here, we provided a comprehensive review of the computational tools and online resources for high-throughput biomedical data analysis. We have discussed computational methods for ScRNA sequencing, multi-omics data integration, AI drug discovery processes, medical imaging data analysis, and the application of IoT in healthcare. We described the fundamental biological terms. Next, we provided the categories of various research problems in health care systems. Third, we demonstrated computational methods for different high-throughput sequencing data (including traditional bulk sequencing profile, single cell sequencing data, micro-array data, drug-cell line interaction data as well as medical imaging data). Finally, we provided different categories of IoT methods/applications for the aforementioned problems.

Acknowledgments

Authors thank Department of Biotechnology (No. BT/RLF/Re-entry/32/2017), Government of India for funding this project.

References

H.J.W.L. Aerts. Data science in radiology: A path forward. *Clinical Cancer Research*, 24(3):532–534, November 2017. doi: 10.1158/1078-0432.ccr-17-2804.

U.D. Akavia, O. Litvin, J. Kim, F. Sanchez-Garcia, D. Kotliar, and et al. An integrated approach to uncover drivers of cancer. *Cell*, 143(6):1005–1017, December 2010. doi: 10.1016/j.cell.2010.11.013.

D.B. Allison, X. Cui, G.P. Page, and M. Sabripour. Microarray data analysis: from disarray to consolidation and consensus. *Nature Reviews Genetics*, 7(1): 55–65, January 2006. doi: 10.1038/nrg1749.

T.S. Andrews and M. Hemberg. M3Drop: dropout-based feature selection for scRNASeq, *Bioinformatics*, 35 (16): 2865–2867, August 2019, https://doi.org/10.1093/bioinformatics/bty1044.

M.R. Aure, I. Steinfeld, L.O. Baumbusch, K. Liestøl, D. Lipson, and et al. Identifying in-trans process associated genes in breast cancer by integrated analysis of copy number and expression data. *PLoS One*, 8(1):e53014, January 2013. doi: 10.1371/journal.pone.0053014.

M. Awale and J.L. Reymond. Polypharmacology browser PPB2: Target prediction combining nearest neighbors with machine learning. *Journal of Chemical Information and Modeling*, 59(1):10–17, December 2018. doi: 10.1021/acs.jcim.8b00524.

S. Bandyopadhyay and S. Mallik. Integrating multiple data sources for combinatorial marker discovery: A study in tumorigenesis. *IEEE/ACM Transactions on Computational Biology and Bioinformatics*, 15(2), December 2016. doi: 10.1109/TCBB.2016.2636207.

S.C. Bendall, K.L. Davis, E.D. Amir, M.D. Tadmor, E.F. Simonds, and et al. Single-cell trajectory detection uncovers progression and regulatory coordination in human B cell development. *Cell*, 157(3):714–725, April 2014. doi: 10.1016/j.cell.2014.04.005.

E.A.G Blomme and Y. Will. Toxicology strategies for drug discovery: Present and future. *Chemical Research in Toxicology*, 29(4):473–504, December 2015. doi: 10.1021/acs.chemrestox.5b00407.

E. Bonnet, L. Calzone, and T. Michoel. Integrative multi-omics module network inference with lemon-tree. *PLoS Computational Biology*, 11(2):e1003983, February 2015. doi: 10.1371/journal.pcbi.1003983.

F. Buettner, K.N. Natarajan, F.P. Casale, V. Proserpio, A. Scialdone, and et al. Computational analysis of cell-to-cell heterogeneity in single-cell RNA-sequencing data reveals hidden subpopulations of cells. *Nature Biotechnology*, 33(2):155–160, January 2015. doi: 10.1038/nbt.3102.

K.R. Campbell and C. Yau. Order under uncertainty: Robust differential expression analysis using probabilistic models for pseudotime inference. *PLoS Computational Biology*, 12(11):e1005212, November 2016a. doi: 10.1371/journal.pcbi.1005212.

K.R. Campbell and C. Yau. switchde: inference of switch-like differential expression along single-cell trajectories. *Bioinformatics*, page btw798, December 2016b. doi: 10.1093/bioinformatics/btw798.

K.R. Campbell and C. Yau. A descriptive marker gene approach to single-cell pseudotime inference. June 2016c. doi: 10.1101/060442.

H.C.S. Chan, J. Wang, K. Palczewski, S. Filipek, H. Vogel, Z.J. Liu, and S. Yuan. Exploring a new ligand binding site of g protein-coupled receptors. *Chemical Science*, 9(31):6480–6489, 2018. doi: 10.1039/c8sc01680a.

H.C.S. Chan, Y. Li, T. Dahoun, H. Vogel, and S. Yuan. New binding sites, new opportunities for GPCR drug discovery. *Trends in Biochemical Sciences*, 44 (4):312–330, April 2019. doi: 10.1016/j.tibs.2018.11.011.

R. Chari, B.P. Coe, E.A. Vucic, W.W. Lockwood, and W.L. Lam. An integrative multi-dimensional genetic and epigenetic strategy to identify aberrant genes and pathways in cancer. *BMC Systems Biology*, 4(1), May 2010. doi: 10.1186/1752-0509-4-67.

J. Chen and S. Zhang. Integrative analysis for identifying joint modular patterns of gene-expression and drug-response data. *Bioinformatics*, 32(11):1724–1732, February 2016. doi: 10.1093/bioinformatics/btw059.

J. Chen, F.D. Bushman, J.D. Lewis, G.D. Wu, and H. Li. Structure-constrained sparse canonical correlation analysis with an application to microbiome data analysis. *Biostatistics*, 14(2):244–258, October 2012. doi: 10.1093/biostatistics/kxs038.

D.Y. Cho and T.M. Przytycka. Dissecting cancer heterogeneity with a probabilistic genotype-phenotype model. *Nucleic Acids Research*, 41(17): 8011–8020, July 2013. doi: 10.1093/nar/gkt577.

C.W. Coley, L. Rogers, W.H. Green, and K.F. Jensen. SCScore: Synthetic complexity learned from a reaction corpus. *Journal of Chemical Information and Modeling*, 58(2):252–261, January 2018. doi: 10.1021/acs.jcim.7b00622.

R.S. Deshmukh, K.A. Kovács, and A. Dinnyés. Drug discovery models and toxicity testing using embryonic and induced pluripotent stem-cell-derived cardiac and neuronal cells. *Stem Cells International*, 2012:1–9, 2012. doi: 10.1155/2012/379569.

J.D. Durrant and J.A. McCammon. NNScore 2.0: A neural-network receptor–ligand scoring function. *Journal of Chemical Information and Modeling*, 51(11):2897–2903, November 2011. doi: 10.1021/ci2003889.

D. Duvenaud, D. Maclaurin, J. Aguilera-Iparraguirre, R. Gómez-Bombarelli, T. Hirzel, and et al. Convolutional networks on graphs for learning molecular fingerprints. In *Proceedings of the 28th International Conference on Neural Information Processing Systems - Volume 2*, NIPS'15, page 2224–2232, Cambridge, MA, USA, 2015. MIT Press. doi: 10.5555/2969442.2969488.

E.N. Feinberg, D. Sur, Z. Wu, B.E. Husic, H. Mai, and et al. PotentialNet for molecular property prediction. *ACS Central Science*, 4(11):1520–1530, November 2018. doi: 10.1021/acscentsci.8b00507.

G. Finak, A. McDavid, M. Yajima, J. Deng, V. Gersuk, and et al. MAST: a flexible statistical framework for assessing transcriptional changes and characterizing heterogeneity in single-cell RNA sequencing data. *Genome Biology*, 16(1), December 2015. doi: 10.1186/s13059-015-0844-5.

D. Grün, A. Lyubimova, L. Kester, K. Wiebrands, O. Basak, and et al. Single-cell messenger RNA sequencing reveals rare intestinal cell types. *Nature*, 525(7568):251–255, August 2015. doi: 10.1038/nature14966.

M. Guo, H. Wang, S.S. Potter, J.A. Whitsett, and Y. Xu. SINCERA: A pipeline for single-cell RNA-seq profiling analysis. *PLoS Computational Biology*, 11 (11):e1004575, November 2015. doi: 10.1371/journal.pcbi.1004575.

Gubern-Merida, Albert & Kallenberg, Michiel & Mann, Ritse & Marti, Robert & Karssemeijer, Nico. (2015). Breast Segmentation and Density Estimation in Breast

MRI: A Fully Automatic Framework. IEEE journal of biomedical and health informatics. 19. 349–57. 10.1109/JBHI.2014.2311163.

L. Haghverdi, M. Buttner, FA. Wolf, F Buettner and FJ Theis. Diffusion pseudotime robustlyreconstructs lineage branching. *Nat Methods*, 2016 (13) 845–848.

T. Ideker, O. Ozier, B. Schwikowski, and A.F. Siegel. Discovering regulatory and signalling circuits in molecular interaction networks. *Bioinformatics*, 18 (Suppl 1):S233–S240, July 2002. doi: 10.1093/bioinformatics/18.suppl_1.s233.

E.M. Jennings, J.S. Morris, R.J. Carroll, G.C. Manyam, and V. Baladandayuthapani. Bayesian methods for expression-based integration of various types of genomics data. *EURASIP Journal on Bioinformatics and Systems Biology*, 2013(1), September 2013. doi: 10.1186/1687-4153-2013-13.

I.C. Jeong, D. Bychkov, and P.C. Searson. Wearable devices for precision medicine and health state monitoring. *IEEE Transactions on Biomedical Engineering*, 66(5):1242–1258, May 2019. doi: 10.1109/tbme.2018.2871638.

Jiao, Han & Jiang, Xinhua & Pang, Zhiyong & Lin, Xiaofeng & Huang, Yihua & Li, Li. (2020). Deep Convolutional Neural Networks-Based Automatic Breast Segmentation and Mass Detection in DCE-MRI. Computational and Mathematical Methods in Medicine. 2020. 1–12. 10.1155/2020/2413706.

P.V Kharchenko, L. Silberstein, and D.T. Scadden. Bayesian approach to single-cell differential expression analysis. *Nature Methods*, 11(7):740–742, May 2014. doi: 10.1038/nmeth.2967.

D. Kim, R. Li, S.M. Dudek, and M.D. Ritchie. ATHENA: Identifying interactions between different levels of genomic data associated with cancer clinical outcomes using grammatical evolution neural network. *BioData Mining*, 6(1), December 2013a. doi: 10.1186/1756-0381-6-23.

D. Kim, R. Li, S.M. Dudek, and M.D. Ritchie. ATHENA: Identifying interactions between different levels of genomic data associated with cancer clinical outcomes using grammatical evolution neural network. *BioData Mining*, 6(1), December 2013b. doi: 10.1186/1756-0381-6-23.

P. Kirk, J.E. Griffin, R.S. Savage, Z. Ghahramani, and D.L. Wild. Bayesian correlated clustering to integrate multiple datasets. *Bioinformatics*, 28(24): 3290–3297, October 2012. doi: 10.1093/bioinformatics/bts595.

V.Y. Kiselev, K. Kirschner, M.T. Schaub, T. Andrews, A. Yiu, and et al. SC3 - consensus clustering of single-cell RNA-seq data. January 2016. doi: 10.1101/036558.

S. kok swee & Chia, F.K. & Nia, M.E. & Tso, C. P. & Chong, A.K. & Abbas, Siti & Chong, S.S. (2014). Breast cancer detection from MR images through an auto-probing discrete Fourier transform system. Computers in biology and medicine. DOI: 10.1016/j.compbiomed.2014.03.00349C. 46–59. 10.1016/j.compbiomed.2014.03.003.

K.D. Korthauer, L.F. Chu, M.A. Newton, Y. Li, J. Thomson, and et al. A statistical approach for identifying differential distributions in single-cell RNA-seq experiments. *Genome Biology*, 17(1), October 2016. doi: 10.1186/s13059-016-1077-y.

D. Krewski, D. Acosta, M. Andersen, H. Anderson, J.C. Bailar, and et al. Toxicity testing in the 21st century: A vision and a strategy. *Journal of Toxicology and Environmental Health*, 13(2–4):51–138, June 2010. doi: 10.1080/10937404.2010.483176.

G.R.G. Lanckriet, T. De Bie, N. Cristianini, M.I. Jordan, and W.S. Noble. A statistical framework for genomic data fusion. *Bioinformatics*, 20(16): 2626–2635, May 2004. doi: 10.1093/bioinformatics/bth294.

J.T. Leek, W.E. Johnson, H.S. Parker, A.E. Jaffe, and J.D. Storey. The sva package for removing batch effects and other unwanted variation in high-throughput experiments. *Bioinformatics*, 28(6):882–883, January 2012. doi: 10.1093/bioinformatics/bts034.

W. Li, S. Zhang, C.C. Liu, and X.J. Zhou. Identifying multi-layer gene regulatory modules from multi-dimensional genomic data. *Bioinformatics*, 28 (19):2458–2466, August 2012. doi: 10.1093/bioinformatics/bts476.

D. Lin, J. Zhang, J. Li, V.D. Calhoun, H.W. Deng, and Y.P. Wang. Group sparse canonical correlation analysis for genomic data integration. *BMC Bioinformatics*, 14(1), August 2013. doi: 10.1186/1471-2105-14-245.

E.F. Lock and D.B. Dunson. Bayesian consensus clustering. *Bioinformatics*, 29 (20):2610–2616, August 2013. doi: 10.1093/bioinformatics/btt425.

T. Lönnberg, V. Svensson, K.R. James, D. Fernandez-Ruiz, I. Sebina, and et al. Single-cell RNA-seq and computational analysis using temporal mixture modeling resolves t h 1/t FH fate bifurcation in malaria. *Science Immunology*, 2(9), March 2017. doi: 10.1126/sciimmunol.aal2192.

R. Louhimo and S. Hautaniemi. CNAmet: an r package for integrating copy number, methylation and expression data. *Bioinformatics*, 27(6):887–888, January 2011. doi: 10.1093/bioinformatics/btr019.

M.I. Love, W. Huber, and S. Anders. Moderated estimation of fold change and dispersion for RNA-seq data with DESeq2. *Genome Biology*, 15(12), December 2014. doi: 10.1186/s13059-014-0550-8.

I.C. Macaulay, V. Svensson, C. Labalette, L. Ferreira, and et al. Single-cell RNA-sequencing reveals a continuous spectrum of differentiation in hematopoietic cells. *Cell Reports*, 14(4):966–977, February 2016. doi: 10.1016/j.celrep.2015.12.082.

P.M. Magwene, P. Lizardi, and J. Kim. Reconstructing the temporal ordering of biological samples using microarray data. *Bioinformatics*, 19(7):842–850, May 2003. doi: 10.1093/bioinformatics/btg081.

A. Mayr, G. Klambauer, T. Unterthiner, and S. Hochreiter. DeepTox: Toxicity prediction using deep learning. *Frontiers in Environmental Science*, 3, February 2016. doi: 10.3389/fenvs.2015.00080.

D.J. McCarthy, K.R. Campbell, A.T.L. Lun, and Q.F. Wills. scater: pre-processing, quality control, normalisation and visualisation of single-cell RNA-seq data in r. August 2016. doi: 10.1101/069633.

Y. Mendelson, D.K. Dao, and K.H. Chon. Multi-channel pulse oximetry for wearable physiological monitoring. In *2013 IEEE International Conference on Body Sensor Networks*, pages 1–6, 2013. doi: 10.1109/BSN.2013.6575518.

R. Miotto, F. Wang, S. Wang, X. Jiang, and J.T. Dudley. Deep learning for healthcare: review, opportunities and challenges. *Briefings in Bioinformatics*, 19(6):1236–1246, May 2017. doi: 10.1093/bib/bbx044.

Q. Mo, S. Wang, V.E. Seshan, A.B. Olshen, N. Schultz, and et al. Pattern discovery and cancer gene identification in integrated cancer genomic data. *Proceedings of the National Academy of Sciences*, 110(11):4245–4250, February 2013a. doi: 10.1073/pnas.1208949110.

Q. Mo, S. Wang, V.E. Seshan, A.B. Olshen, N. Schultz, and et al. Pattern discovery and cancer gene identification in integrated cancer genomic data. 110(11):4245–4250, February 2013b. doi: 10.1073/pnas.1208949110.

M. Olivecrona, T. Blaschke, O. Engkvist, and H. Chen. Molecular de-novo design through deep reinforcement learning. *Journal of Cheminformatics*, 9 (1), September 2017. doi: 10.1186/s13321-017-0235-x.

K. Ovaska, M. Laakso, S. Haapa-Paananen, R. Louhimo, P. Chen, and et al. Large-scale data integration framework provides a comprehensive view on glioblastoma multiforme. *Genome Medicine*, 2(9):65, 2010. doi: 10.1186/gm186.

K. Pearson. LIII. on lines and planes of closest fit to systems of points in space. *The London, Edinburgh, and Dublin Philosophical Magazine and Journal of Science*, 2(11):559–572, November 1901. doi: 10.1080/14786440109462720.

E. Pierson and C. Yau. ZIFA: Dimensionality reduction for zero-inflated single-cell gene expression analysis. *Genome Biology*, 16(1), November 2015. doi: 10.1186/s13059-015-0805-z.

X. Qiu, Q. Mao, Y. Tang, L. Wang, R. Chawla, and et al. Reversed graph embedding resolves complex single-cell developmental trajectories. February 2017. doi: 10.1101/110668.

B. Ramsundar, P. Eastman, P. Walters, and V. Pande. O'Reilly Media, April 2019.

P. Ray, L. Zheng, J. Lucas, and L. Carin. Bayesian joint analysis of heterogeneous genomics data. *Bioinformatics*, 30(10):1370–1376, January 2014. doi: 10.1093/bioinformatics/btu064.

M.D. Robinson, D.J. McCarthy, and G.K. Smyth. edgeR: a bioconductor package for differential expression analysis of digital gene expression data. *Bioinformatics*, 26(1):139–140, November 2009. doi: 10.1093/bioinformatics/btp616.

M. Ruffalo, M. Koyutürk, and R. Sharan. Network-based integration of disparate omic data to identify "silent players" in cancer. *PLoS Computational Biology*, 11(12):e1004595, December 2015. doi: 10.1371/journal.pcbi.1004595.

B. Sanchez-Lengeling, C. Outeiral, G.L. Guimaraes, and A. Aspuru-Guzik. Optimizing distributions over molecular space. an objective-reinforced generative adversarial network for inverse-design chemistry (ORGANIC). 2017. doi: 10.26434/chemrxiv.5309668.v3.

J. Sander, J.L. Schultze, and N. Yosef. ImpulseDE: detection of differentially expressed genes in time series data using impulse models. *Bioinformatics*, page btw665, October 2016. doi: 10.1093/bioinformatics/btw665.

J.A. Seoane, I.N.M. Day, T.R. Gaunt, and C. Campbell. A pathway-based data integration framework for prediction of disease progression. *Bioinformatics*, 30(6):838–845, October 2013. doi: 10.1093/bioinformatics/btt610.

M. Setty, M.D. Tadmor, S. Reich-Zeliger, O. Angel, T.M. Salame, and et al. Wishbone identifies bifurcating developmental trajectories from single-cell data. *Nature Biotechnology*, 34(6):637–645, May 2016. doi: 10.1038/nbt.3569.

R. Shen, Q. Mo, N. Schultz, V.E. Seshan, A.B. Olshen, and et al. Integrative subtype discovery in glioblastoma using iCluster. *PLoS One*, 7(4):e35236, April 2012. doi: 10.1371/journal.pone.0035236.

N.K. Speicher and N. Pfeifer. Integrating different data types by regularized unsupervised multiple kernel learning with application to cancer subtype discovery. *Bioinformatics*, 31(12):i268–i275, June 2015. doi: 10.1093/bioinformatics/btv244.

T. Si, A. De, A. Kumar. Artificial Neural Network based Lesion Segmentation of Brain MRI. 4(5) February 2016, Communications on Applied Electronics (CAE) - ISSN : 2394 – 4714, DOI: 10.5120/cae2016652096.

S. Steiner, J. Wolf, S. Glatzel, A. Andreou, J.M. Granda, and et al. Organic synthesis in a modular robotic system driven by a chemical programming language. *Science*, 363(6423):eaav2211, November 2018. doi: 10.1126/science.aav2211.

C. Stork, Y. Chen, M. Šícho, and J. Kirchmair. Hit dexter 2.0: Machine-learning models for the prediction of frequent hitters. *Journal of Chemical Information and Modeling*, 59(3):1030–1043, January 2019. doi: 10.1021/acs.jcim.8b00677.

C.A. Vallejos, J.C. Marioni, and S. Richardson. BASiCS: Bayesian analysis of single-cell sequencing data. *PLoS Computational Biology*, 11(6):e1004333, June 2015. doi: 10.1371/journal.pcbi.1004333.

L. van der Maaten and G. Hinton. Visualizing data using t-sne. *Journal of Machine Learning Research*, 9(86):2579–2605, 2008. accessed on day month year.

C.J. Vaske, S.C. Benz, J.Z. Sanborn, D. Earl, C. Szeto, and et al. Inference of patient-specific pathway activities from multi-dimensional cancer genomics data using PARADIGM. *Bioinformatics*, 26(12):i237–i245, June 2010. doi: 10.1093/bioinformatics/btq182.

B. Wang, A.M. Mezlini, F. Demir, M. Fiume, Z. Tu, and et al. Similarity network fusion for aggregating data types on a genomic scale. *Nature Methods*, 11(3): 333–337, January 2014. doi: 10.1038/nmeth.2810.

Y. Wang, G. Morrell, M.E. Heibrun, A. Payne, D.L. Parker. 3D multi-parametric breast MRI segmentation using hierarchical support vector machine with coil sensitivity correction. Acad Radiol. 2013;20(2):137–147. doi:10.1016/j.acra.2012.08.016.

B. Wang, J. Zhu, E. Pierson, D. Ramazzotti, and S. Batzoglou. Visualization and analysis of single-cell RNA-seq data by kernel-based similarity learning. *Nature Methods*, 14(4):414–416, March 2017. doi: 10.1038/nmeth.4207.

G. Wang. A perspective on deep imaging. *IEEE Access*, 4:8914–8924, 2016. doi: 10.1109/ACCESS.2016.2624938.

M. Wójcikowski, P. Zielenkiewicz, and P. Siedlecki. Open drug discovery toolkit (ODDT): a new open-source player in the drug discovery field. *Journal of Cheminformatics*, 7(1), June 2015. doi: 10.1186/s13321-015-0078-2.

C. Xu and Z. Su. Identification of cell types from single-cell transcriptomes using a novel clustering method. *Bioinformatics*, 31(12):1974–1980, February 2015. doi: 10.1093/bioinformatics/btv088.

Y. Xu, J. Ma, A. Liaw, R.P. Sheridan, and V. Svetnik. Demystifying multitask deep neural networks for quantitative structure–activity relationships. *Journal of Chemical Information and Modeling*, 57(10):2490–2504, October 2017. doi: 10.1021/acs.jcim.7b00087.

N. Yasuo and M. Sekijima. Improved method of structure-based virtual screening via interaction-energy-based learning. *Journal of Chemical Information and Modeling*, 59(3):1050–1061, February 2019. doi: 10.1021/acs.jcim.8b00673.

Z.H. You, Z. Yin, K. Han, D.S. Huang, and X. Zhou. A semi-supervised learning approach to predict synthetic genetic interactions by combining functional and topological properties of functional gene network. *BMC Bioinformatics*, 11(1), June 2010. doi: 10.1186/1471-2105-11-343.

Y. Yuan, R.S. Savage, and F. Markowetz. Patient-specific data fusion defines prognostic cancer subtypes. *PLoS Computational Biology*, 7(10):e1002227, October 2011. doi: 10.1371/journal.pcbi.1002227.

S. Zhang, Q. Li, J. Liu, and X.J. Zhou. A novel computational framework for simultaneous integration of multiple types of genomic data to identify microRNA-gene regulatory modules. *Bioinformatics*, 27(13):i401–i409, June 2011. doi: 10.1093/bioinformatics/btr206.

K. Zhou., H. Greenspan, and D. Shen. Academic Press, 2017.

6

Association Rule Mining Based on Ethnic Groups and Classification using Super Learning[1]

Md Faisal Kabir[1,] and Simone A. Ludwig[2]*

[1]*Department of Computer Science, Pennsylvania State University-Harrisburg, 17057, PA, Middletown, 777 West Harrisburg Pike, USA*
[2]*Department of Computer Science, North Dakota State University, 58102, ND, Fargo, 1320 Albrecht Blvd, USA*

6.1 Introduction

Cancer is one of the deadliest diseases worldwide. According to the World Health Organization (WHO) (Ferlay et al., 2015), there are more than 10 million new cases reported each year. Cancer affects nearly every household, although cancer types are prevalent in varying geographical areas. One example is breast cancer, which is the most prevalent type of cancer in women worldwide, with 1.7 million new cases diagnosed in 2012 (Ferlay et al., 2015). Therefore, prevention strategies are needed to address this issue. Identifying the risk factors of breast cancer is crucial, since it allows physicians to consult with their patients about the associated risks. Accordingly, physicians can recommend precautionary actions.

Data mining can be described as extracting implicit, concealed, and valuable knowledge from a massive amount of data (Han et al., 2001). Data mining encompasses several different techniques. Rule mining is a method provides mining knowledge in the form of rules, which are easily understood (Rahman et al., 2012). Association rule mining (ARM) (Agrawal et al., 1993) is a particular category of rule mining that was introduced in 1993 and widely used in industry as a "market-basket analysis". The goal is to discover knowledge in the form of rules that tracks customer buying patterns. By applying this systemic method, some items can be inferred, given the appearance of other items in a transaction (Aggarwal and Yu, 1998). Such knowledge is beneficial for decisions such as buyer targeting, shelving, and advertisements. In a market-basket analysis, one rule could be ((item = milk, item = sugar, item = bread) =>(item = butter)). This rule indicates that if a customer purchases milk, sugar, and bread, it is most probable that the customer will likewise buy butter. The association rule technique has also been applied to several domains,

*Corresponding Author: Md Faisal Kabir; mpk5904@psu.edu
[1]Analysis of Breast Cancer Risk Factor Data.

Applied Smart Health Care Informatics: A Computational Intelligence Perspective, First Edition.
Edited by Sourav De, Rik Das, Siddhartha Bhattacharyya, and Ujjwal Maulik.

in particular the medical domain (Kabir et al., 2018; Khalilian and Tabibi, 2015; Ordonez et al., 2000; Stilou et al., 2001).

We will first discuss the discovery or significant rules for breast cancer patients with a focus on ethnic groups. Foretelling the risk of breast cancer is an essential issue for clinical oncologists. A reliable prediction will help oncologists and other clinicians in their decision-making process and allow doctors to choose the most reliable and evidence-based therapy. Moreover, the best prevention procedures for patients can be identified.

Classification is a supervised learning technique that groups unknown or test data into a finite set of classes by learning an objective function that maps independent attributes into one of the target classes (Kabir and Ludwig, 2019; Agrawal et al., 1992; Rahman et al., 2014). The objective function is referred to as the classification model. Classification is applied to many different fields with the aim of generating the best performing model by experimenting with different classification algorithms. The usual procedure for achieving better performance of a data set is to use a single classifier. However, a single classifier usually does not provide the best performance, so researchers have looked at different techniques to address this. For example, multiple models could be used for the classification task. Researchers have been using Bagging and Boosting ensemble methods in many areas to obtain a better performance (Kaur and Batra, 2017; Gibbons et al., 2017; Silwattananusarn et al., 2016). Lately, a technique called super-learning or stacking has been introduced that aggregates several algorithms to obtain a better predictive model (Kabir and Ludwig, 2019; Laan; Van der Laan and Rose, 2011).

Second, we will apply a super-learning technique to the data set. The super learner uses three machine learning algorithms, namely gradient boosting machine (GBM), random forest (RF), and deep neural network (DNN). The generalized linear model (GLM) is used as the meta-learner (Vanerio and Casas, 2017; Aiello et al., 2016). Performances of the super learner and the individual base learners were compared.

After the introduction 6.1, we will describe the background or related work for this study in 6.2. The motivation and contribution of our work are outlined in 6.3. The description of the data used for this research is stated in 6.4 and the methodology in 6.5, where we demonstrate our proposed approaches. In 6.6, we describe our experiments. In particular, the evaluation criteria of the classification model and the super learner's results are shown and presented along with a discussion 6.6.3, where we summarize our studies and limitations. In 6.7 is a summary to conclude our findings and present future work directions.

6.2 Background

Researchers have investigated breast cancer risk factors to find relationships among them; they have also developed various breast cancer risk prediction models (Hou et al., 2013; Gail et al., 1989; Barlow et al., 2006; Gauthier et al., 2011). Hou et al. (2013), used statistical methods to examine the relationship between hormone replacement therapy (HRT) and breast cancer risk, and they found that

HRT increased the risk of breast cancer. Gail et al. (1989) used the Gali model to assess the number of breast cancer cases for white women examined annually. Barlow et al. (2006) used identified breast cancer risk factors to describe the model. Furthermore, a data mining approach named k-nearest-neighbor (KNN) is applied to determine the breast cancer risk score that ultimately improves the readability for physicians and patients (Gauthier et al., 2011).

A data mining technique called association rule mining (ARM) has been applied to the medical field to extract knowledge in the form of rules from the data. Khalilian and Tabibi (2015) used the ARM-based technique to find co-occurrences of infections carried by a sufferer from a healthcare database. The method collected data from a patients' healthcare repository from which association rules were discovered. The researcher also investigated class ARM, a variation of the ARM technique, to discover the characteristic features (Li et al., 2001). By definition, an ARM is a subset of association rules with the particular classes as their consequences (Paul et al., 2014a). If we assign a very low support value in traditional ARM, then the class ARM will generate overfitting rules for a frequent class. On the other hand, if we specify a very high support value, then an insufficient number of rules for an infrequent class will be generated. This issue can be avoided in the class ARM as mining is performed based on the target class, and thus, the algorithm is not affected by the unbalanced proportion between the classes. As an example, Kabir et al. (2018) produced valuable rules from risk factor data for both breast cancer and non-breast cancer patients.

There are ethnic and/or racial inequalities in cancer occurrence and survival. Racial differences in cancer can include socioeconomic standing and health care access. Ethnicity concerns the different cultural, socioeconomic, and religious properties including customs, language, diet, and cultural identity (abu), all of which are linked to lifestyle and behavioral factors that can lead to cancer disparities by ethnic status (Gupta et al., 2012). As physically similar ethicities are often socially grouped into races, genetic differences due to ancestry may appear to be racially linked. Cancer survival can also be affected at the epigenetic level due to the distinct phenotypic features, drug metabolism, and disease susceptibility (Heyn et al., 2013). For example, triple-negative breast cancer (ER-, PR-, HER2-), one of the most disruptive forms of breast cancer, is significantly more prevalent in Black women (20 to 50%) than other races and/or ethnicities (9 to 15%) (Agboola et al., 2012; Clarke et al., 2012). Genetic factors also contribute to differences in the cancer susceptibility of the Black population compared to white people in the US (Yao et al., 2013).

In the first part of this chapter, we discovered hidden but significant rules for people with breast cancer based on their ethnic group. Rules of breast cancer patients from different ethnic groups can help physicians decide and familiarize patients about risk factors. Also, physicians can warn patients regarding the possible hazards of developing breast cancer. In this way, a prevention plan for particular ethnicities can start in the initial disease progression step.

Machine Learning (ML) techniques have been applied in the medical field to inform decision-making processes. For instance, to predict cancer risk. Kabir and Ludwig (2018) applied classification methods. The authors also used various resampling approaches on the training data as risk factor data often have an unequal proportion of cancer and non-cancer cases.

Ensemble techniques have been applied in different fields, including the medical field, to obtain more reliable predictive performance (Kaur and Batra, 2017; Gibbons et al., 2017; Silwattananusarn et al., 2016). Researchers also use a super learning method that aggregates a group of learning algorithms (Laan; Van der Laan and Rose, 2011). Kabir and Ludwig (2019) used two different forms of super learner (SL); the first SL consisted of two base learners, and the other of three algorithms. The authors showed that the SL with three base learners performed better than the one with two base learners and all individual ML algorithms that they applied for their research. The authors used four popular data sets to assess the performance of their techniques.

In the second part of this chapter, we present a SL technique with three ML algorithms, specifically GBM, RF, and DNN; GLM was selected as a meta-learner (Vanerio and Casas, 2017; Aiello et al., 2016). We compare the performance of the SL with the individual base learners; it showed that our SL method provides satisfactory results over the individual algorithms considered.

6.3 Motivation and Contribution

Breast cancer is the most deadly type of cancer in women worldwide; 1.7 million new cases were diagnosed in 2012 (Ferlay et al., 2015). Preventing breast cancer through a quantified risk assessment is of significant concern to lessen its impact on society. Identifying breast cancer risk factors to reduce ethnicity- or race-based disparities is essential. Physicians can then familiarize patients with the relevant risk factors and suggest preventive measures. It is also crucial to extract meaningful knowledge from these risk factors in a concise form like rules. The obtained rules can be beneficial for better healthcare as medical specialists or other health-related organizations can develop policies to identify and prevent its impact early. For that reason, this research applied class ARM and discovered hidden but significant rules for breast cancer patients of different ethnic groups.

Classification is a common task in the machine learning domain and has been widely applied to diverse application domains including healthcare. However, learning algorithms' performance heavily depends on sufficient qualitative training data to build an accurate model and predict future or unseen data. Nonetheless, in real-life settings, especially in the healthcare area, obtaining useful training data has been a significant challenge in making efficient prediction models that can be applied in practice. It is more important to understand each problem and select the appropriate technique carefully to overcome these issues. Researchers are also using multiple models for a particular problem to obtain better performance. Super learning is an ML technique that finds the optimal weighted average of diverse learning models and usually provides a better performance than the individual base learners. For this purpose, enhancing the performance of classification with super learning was developed.

This chapter makes a few contributions towards better healthcare using data mining and machine learning methods. Useful knowledge in the form of rules using DM techniques have been discovered as breast cancer risk factors of ethnic groups. These rules will be useful to understand and compare the characteristics of ethnicities as it relates to breast cancer. Moreover, building predictive models utilizing the SL technique has been investigated to enhance classification models' accuracy. The performance of the proposed technique was compared to the individual machine learning algorithms.

6.4 Data Analysis

6.4.1 Data Description

The data set comprises information from 6 318 638 mammography examinations that were collected from the Breast Cancer Surveillance Consortium (BCSC) database. The data-gathering period ran from January 2000 to December 2009. More information about BCSC data resources can be found at bcsc-research.org.

6.4.2 Data Preprocessing

In the BCSC risk factor data set, the total number of cases is 1 144 565, with 13 attributes or features. For simplicity, we dismissed records having at least one missing value, which was denoted by 9 in the data set. We also removed the calendar year of the observation. After discarding these instances and column, there are 219 524 usable records with 12 features. Detailed preprocess steps of the BCSC risk factor data set are illustrated in Kabir et al. (2018). After further preprocessing, there were 11 attributes with 1 015 583 instances. Among these, the number of breast cancer patients and non-breast cancer patients are 60 800 and 954 783, respectively. Attribute distributions for the risk factor data can be found in Kabir et al. (2018).

6.4.3 Further Preprocessing for Ethnic Group Rule Discovery with Multiple Consequences

Our goal is to extract hidden, but useful, information in the form of rules for breast cancer occurrence across ethnic groups from the risk factor data set. For that, we merged two attributes named breast cancer history (where the value is Yes - meaning we are considering breast cancer patients) and patient race/ethnicity attribute (race-cancer-history). The distribution for this combined attribute group is shown in Table 6.1. For instance, the Non-Hispanic-White-Yes value of the attribute race/ethnicity represents breast cancer patients in the non-Hispanic white group.

6.4.3.1 Transaction-Like Database for Association Rule

For association and class ARM, the data set is changed into a transaction-like database. For example, for a feature such as BMI_group there were a total of four

Table 6.1 Distribution of breast cancer patients based on race or ethnicity.

Race or ethnicity	Number of individuals
Non-Hispanic-White-Yes	54869
Asian_or_Pacific Islander-Yes	1867
Hispanic-Yes	2028
Other_or_Mixed-Yes	1055
Non-Hispanic-Black-Yes	736
Native-American-Yes	245

values, namely 10 to 24, 25 to 29, 30 to 34, and 35+. These four attributes were produced subsequently with values "Yes" or "No". For example, if an individual's BMI_group (say 22.58) lies within the range of 10 to 24, then the value of the corresponding column (10 to 24) would be "Yes"; the values in the rest of the BMI columns would be "No". The process continues for other attributes, and in sum, 44 columns are produced. So, after converting the data set into a transaction-like database, there were 44 items and 60 800 instances for class association rule mining.

6.4.4 Classification Data Set

For the classification model, we considered both breast cancer and non-breast cancer patients. As discussed in the preprocessing data section, there are 60 800 breast cancer and 954 783 non-breast cancer patients with 11 attributes. Amidst those 11 columns, a "prior breast cancer" value of "Yes" or "No" is considered a response, or class attribute; the remaining 10 features are recognized as predictors. For classification, 80% of the data was selected for the training, and the remaining 20% kept as a test set (Kabir and Ludwig, 2018). The stratified shuffle split method from (sklearn) was applied as it conserves the proportion of samples for each group, which is vital for imbalanced data. It is worth mentioning that sci-kit-learn is a machine learning library for the Python programming language.

The total number of training examples were 812 466 with 48 640 breast cancer and 763 826 non-breast cancer patients. The total number of test instances were 203 117 having 12 160 breast cancer patients and 190 957 non-breast cancer individuals.

The risk factor data have imbalanced characteristics, which indicate that the data have an uneven distribution between the cancer and non-cancer patients; for that reason, the training data has been modified with multiple resampling techniques (Kabir and Ludwig, 2018). Resampling techniques change the data to adjust class distributions before using it to train the classifier. The process is usually performed by eliminating some instances from the majority class or adding more to the minority class. For the classification task in this study, we selected training data that has been modified using the synthetic minority over-sampling technique (SMOTE) and

Table 6.2 Training data obtained by applying SMOTE and ENN.

Resampling technique	Class = yes	Class = no	Total instances
SMOTE + ENN	437 256	658 167	1 095 423

then reduced using the nearest neighbor (ENN) method. SMOTE generates synthetic minority examples by linear interpolation between neighbors in the input space to adjust the class distribution. ENN is a procedure of under-sampling the majority class. It eliminates examples whose class labels deviate from most of its k nearest neighbors (More, 2016). The distribution of the training data that were obtained by applying SMOTE followed by ENN is shown in Table 6.2.

6.5 Methodology

In this section, we first used class ARM on modified risk factor data, discussed in 6.4.3 and 6.4.3.1 to extract useful rules for breast cancer from across ethnic groups. After that, we applied the ensemble technique (SL) to the BCSC risk factor data as discussed in 6.4.4.

6.5.1 Association Rule Mining

ARM is an important technique for generating and discovering knowledge from a large database. Detailed information on generating association rules along with important measures can be found in Han et al. (2001) and Kabir et al. (2018). Rules can be discovered from data sets by specifying particular classes as their consequence, which is a class association rule technique.

The purpose of a class ARM is to apply exhaustive exploration procedures to attaining rules with the target variable as their outcomes that satisfy minimum support and confidence (Paul et al., 2014b). The appropriate value of minimum support and confidence is essential for producing rules; having a value that is too low can generate many rules; using a very high value, however, we may miss rare but valuable rules. More information about class association rule techniques and their usage is available in Kabir et al. (2018) and Paul et al. (2014b). In this research, we applied class ARM and discovered significant rules for breast cancer patients belonging to different ethnic groups. Note that here we have extracted rules with two attributes/items in the consequent at the same time. We merged two attributes into one, namely race-cancer-history, that indicates breast cancer patients of a particular ethnic group; discussed in 6.4.3. As we consider non-Hispanic white, Hispanic, and Asian/Pacific Islander in the class association rule mining process, we ran the algorithm with the consequent value as any of these three values along with specified support and confidence values. For instance, during rules generation for the non-Hispanic white group, we set the consequent as "race-cancer-history = non-Hispanic-White" along with other important measures.

Rules for breast cancer patients from different ethnic groups can be useful for physicians or primary care providers to make decisions. An ethnicity-based prevention plan or process can begin in the initial stage of disease progression.

6.5.2 Super Learning

SL or stacked ensemble generally consists of two or more machine learning (ML) techniques. It is a cross-validation-based technique for incorporating ML methods that generally provide better performance than those of the base algorithm (Laan; Van der Laan and Rose, 2011). The first step of the SL technique is to specify base learners or ML algorithms (say L base learners) and a meta-learning algorithm. The next step is to train the ensemble, which has the following main functions:

- Train all of the defined ML algorithms, also called base learners (L), on the training data.
- Perform K-fold cross-validation (CV) on each of these base learners or classification algorithms and collect the cross-validated predicted values from the K-fold CV conducted on each of the L learners.
- From each L, the M cross-validated predicted values are collected and merged to form a new $M X L$ matrix. The matrix and the real class vector are called the level-1 data. Here, M indicates the number of samples or examples in the training data.
- Train the meta-learning algorithm on the level-1 data. The ensemble-model includes the L classification algorithms and a meta-learning model, which can then be used to generate predictions on a test set or new samples.

The third and final step is to predict new or unknown data. For that, predictions from all base learners are produced and fed into the meta-learner.

Detailed information and applications of SLs can be found in Laan, Kabir and Ludwig (2019), Nykodym et al. (2016), and LeDell (2016). A concept diagram of the SL approach for this research is shown in Figure 6.1.

6.5.2.1 Ensemble or Super Learner Set-Up

For this research, first of all, three well-known machine learning algorithms were specified as base learners: gradient boosting machine (GBM), random forest (RF),

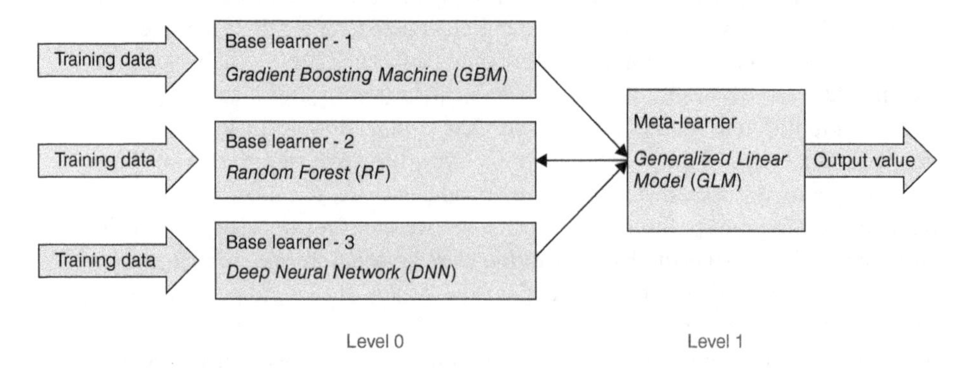

Figure 6.1 Super learner concept diagram (Kabir and Ludwig, 2019).

Table 6.3 Default parameter values for corresponding base learners.

Base learner	Hyper-parameter default values
GBM	learn_rate: [0.1] sample_rate: [1.0] col_sample_rate_per_tree: [1.0] max_depth: [5]
RF	sample_rate: [0.63] col_sample_rate_per_tree: [1.0] max_depth: [20]
DNN	activation: [rectifier] hidden: [200 200] l1: [0.0] l2: [0.0]

and deep neural network (DNN) (Vanerio and Casas, 2017). The generalized linear model (GLM) was selected as a meta-learner (Vanerio and Casas, 2017; Aiello et al., 2016). After that, we trained all individual ML techniques, namely GBM, RF, and DNN. On each of these learning algorithms, the default parameters available in H2O were used (Aiello et al., 2016). A 10-fold CV was done on each of these algorithms; the CV prediction parameter was defined as "True". The target column or the class value for risk factor data was binary so the Bernoulli distribution was selected. From the above three base learners, the cross-validated predicted values were collected from 1 015 583 training samples to form a $1\,015\,583 \times 3$ matrix. This matrix and the real class vector of the training data are referred to as level-1 data for our research.

Following that, we trained the meta-learning algorithm named GLM on the level-1 data. The ensemble-model included the three classification algorithms and the meta-learner to produce estimates on a test set or new samples.

In Table 6.3 the important parameters (default values) for each base learner are listed.

6.6 Experiments and Results

Results obtained using a class association rule are presented in this section. Crucial rules for breast cancer patients according to ethnicity are produced by selecting a suitable value of support and confidence. The interpretations of a few crucial rules are explained in this section – as are the evaluation measures for the classification model and the results obtained from SL. vc For data preprocessing, we applied Java and Python programming languages. For class association rule mining, we used the "arules" package available in R programming. For the classification task, we used

H2O, which is considered the leading open-source ML and AI platform (Aiello et al., 2016). The Python programming language was used for the implementation in H2O.

6.6.1 Rules Discovery

Our objective is to generate characteristics of patients as a form of rules from a particular race/ethnic background with prior breast cancer. For this, we discovered rules applying the class ARM technique with the specified support and confidence. We also marked the consequence of a rule (Race:cancer_history) to get our target rules that depict the individual who has breast cancer and a particular race/ethnic background. From the distribution of breast cancer patients in the race/ethnic groups shown in Table 6.1, we can see that there are very few breast cancer patients of native-American, non-Hispanic-Black, and other-or-mixed race/ethnicity in the BCSC risk factor data set; these groups were eliminated from the rule mining process. In the rule mining process, we consider breast cancer patients for the non-Hispanic white group as well as Hispanic and Asian/Pacific islander.

6.6.1.1 Rules of Breast Cancer Patients Based on Ethnic Groups

After several experiments, the support and confidence values were assigned to 0.005% and 80%, respectively, and we obtained five rules. Here, we specified the consequent value "Asian_or_pacific_Islander_Yes", to obtain the rules of breast cancer patients having Asian_or_pacific_Islander ethnic group. These rules are shown in Table 6.4 while the scatter plot of these rules sorted by lift value is shown in Figure 6.2.

For breast cancer patients in the Hispanic group, after several experiments, the support and confidence values were specified to 0.005% and 85%, respectively, and we obtained five rules. Here, we assign the consequent "Hispanic_Yes" to obtain breast cancer patients' rules belonging to the Hispanic ethnic group. These rules are displayed in Table 6.5, while the scatter plot of those rules is illustrated in Figure 6.3.

For breast cancer patients of the non-Hispanic white group, the assigned support and confidence values were 30% and 90%, respectively; we obtained 23 rules. Here, we specified the consequent or class value as "Non-Hispanic-White-Yes" to obtain the rules of breast cancer patients in the non-Hispanic white group. The scatter plot of these 23 rules are shown in Figure 6.4, while the top five rules sorted by the lift value are shown in Table 6.6.

6.6.1.2 Interpreting Rules

We can interpret rule 1 in Table 6.4 as "If a person's age is between 30 and 34 with no breast cancer in first degree relatives, their first menstrual cycle occurred before the age of 12 years, they have not had children, their breast tissue is heterogeneously dense, and their BMI is between 10 and 25, then an individual belonging to the Asian/Pacific islander race/ethnicity group may be a breast cancer patient".

Rule 1 in Table 6.5 can be interpreted as "If a person's age is between 18 and 29, they had their first child between 20 and 24 years old, and their BMI is between 25

Table 6.4 Discovered rules using the class association rule method (consequent = Asian-or-pacific-Islander-Yes) with corresponding support and confidence.

SL	Rules	Supp. (%)	Conf. (%)	Lift
1	{Age_group=age_30_34, First_degree_relative=No, Age_menarche=Age_less_12, Age_first_birth=Nulliparous, BIRADS_breast_density=Heterogeneously_dense, BMI_group=10-to-lessThan_25} =>{Race_cancer_history=Asian_or_pacific_Islander_Yes}	0.005	100	32.57
2	{Age_group=age_40_44, First_degree_relative=No, Age_first_birth=Age_greater_equal_30, biopsy=No} =>{Race_cancer_history=Asian_or_pacific_Islander_Yes}	0.005	80	26.05
3	{Age_group=age_30_34, Age_menarche=Age_less_12, Age_first_birth=Nulliparous, BIRADS_breast_density=Heterogeneously_dense, BMI_group=10-to-lessThan_25} =>{Race_cancer_history=Asian_or_pacific_Islander_Yes}	0.005	80	26.05
4	{Age_group=age_30_34, Age_menarche=Age_less_12, First_degree_relative=No, Age_first_birth=Nulliparous, BIRADS_breast_density=Heterogeneously_dense} =>{Race_cancer_history=Asian_or_pacific_Islander_Yes}	0.005	80	26.05
5	{First_degree_relative=No, HRT=No, Age_menarche=Age_greaterEqual_14, Age_first_birth=Age_greater_equal_30, Menopaus=post_menopausal, biopsy=No, BMI_group=10-to-lessThan_25} =>{Race_cancer_history=Asian_or_pacific_Islander_Yes}	0.005	80	26.05

and 30, then there is a very high chance that a Hispanic person could have breast cancer".

Similarly, we can describe rule 1 in Table 6.6 as "If a individual's breast density is scattered fibroglandular dense, post-menopausal with no records of hormone replacement therapy, and no previous breast cancer biopsy, then a non-Hispanic white person may be a breast cancer patient".

6.6.2 Evaluation Criteria of Classification Model

Several evaluation metrics were considered, like accuracy, precision, sensitivity/recall, and specificity (Fawcett, 2006) to measure the super learner's

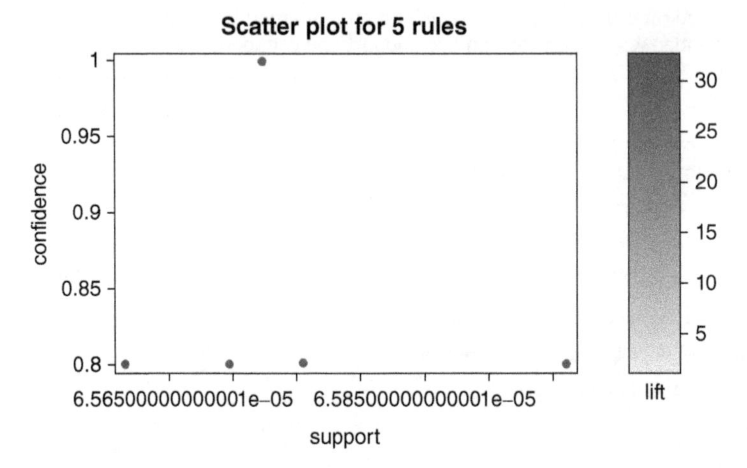

Figure 6.2 Scatter plot of the five rules for breast cancer patients in the Asian/Pacific islander group with specified support, confidence, and lift values.

Table 6.5 Extracted rules using the class association rule technique (consequent = Hispanic-Yes) with corresponding support and confidence value.

SL	Rules	Supp. (%)	Conf. (%)	Lift
1	{Age_group=age_18_29, Age_first_birth=Age_20_24, BMI_group=25-to-lessThan_30} =>{Race_cancer_history=Hispanic_Yes}	0.005	100	29.98
2	{Age_group=age_30_34, Age_first_birth=Age_20_24, BIRADS_breast_density=Extremly_dense} =>{Race_cancer_history=Hispanic_Yes}	0.005	100	29.98
3	{Age_group=age_18_29, Age_menarche=Age_less_12, BMI_group=10-to-lessThan_25, BIRADS_breast_density=Heterogeneously_dense} =>{Race_cancer_history=Hispanic_Yes}	0.005	100	29.98
4	{Age_group=age_45_49, First_degree_relative=No, Age_first_birth=Age_less_20, HRT=Yes, BMI_group=10-to-lessThan_25, BIRADS_breast_density=Heterogeneously_dense} =>{Race_cancer_history=Hispanic_Yes}	0.005	87.5	26.23
5	{Age_group=age_65_69, First_degree_relative=Yes, Age_menarche=Age_greaterEqual_14, Age_first_birth=Age_less_20, HRT=No, biopsy=Yes, BMI_group=10-to-lessThan_25, BIRADS_breast_density=scattered_fibroglandular_densities} =>{Race_cancer_history=Hispanic_Yes}	0.005	87.5	26.23

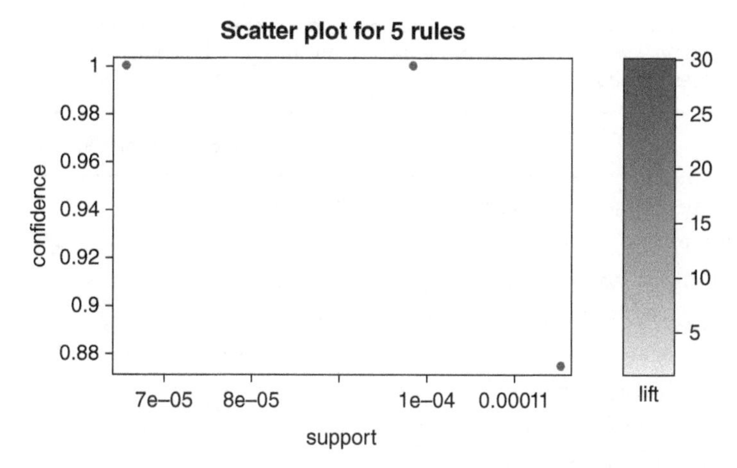

Figure 6.3 Scatter plot of five rules for individuals with breast cancer from the Hispanic group with corresponding support, confidence, and lift values.

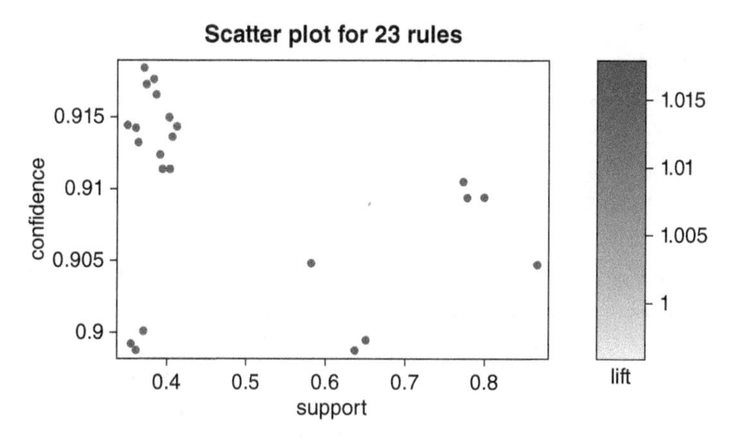

Figure 6.4 Scatter plot of 23 rules for breast cancer patients in the non-Hispanic white group with specified support, confidence, and lift values.

performance. These metrics can be computed with a confusion matrix and employed to assess the model; see Equations 6.1 through 6.4.

$$Accuracy = \frac{\mathrm{TP} + \mathrm{TN}}{\mathrm{TP} + \mathrm{FP} + \mathrm{TN} + \mathrm{FN}} \tag{6.1}$$

$$Sensitivity/Recall = \frac{\mathrm{TP}}{\mathrm{TP} + \mathrm{FN}} \tag{6.2}$$

$$Specificity = \frac{\mathrm{TN}}{\mathrm{TN} + \mathrm{FP}} \tag{6.3}$$

$$Precision = \frac{\mathrm{TP}}{\mathrm{TP} + \mathrm{FP}} \tag{6.4}$$

Here, TP is the number of positive instances that are classified as positive, TN is the number of negative samples accurately classified as negative, FN is the number

Table 6.6 Rules generated using the class association rule technique (consequent set to "Non-Hispanic-White-Yes") with corresponding support, confidence, and lift value. Top 5 rules sort by lift values are shown.

SL	Rules	Supp. (%)	Conf. (%)	Lift
1	{BIRADS_breast_density=scattered_fibroglandular_densities, HRT=No, Menopaus=post_menopausal, biopsy=Yes } =>{Race_cancer_history=Non-Hispanic-White_Yes}	37	92	1.02
2	{BIRADS_breast_density=scattered_fibroglandular_densities, Menopaus=post_menopausal, biopsy=Yes } =>{Race_cancer_history=Non-Hispanic-White_Yes}	38	92	1.02
3	{BIRADS_breast_density=scattered_fibroglandular_densities, HRT=No, Menopaus=post_menopausal } =>{Race_cancer_history=Non-Hispanic-White_Yes}	38	92	1.02
4	{BIRADS_breast_density=scattered_fibroglandular_densities, Menopaus=post_menopausal } =>{Race_cancer_history=Non-Hispanic-White_Yes}	39	92	1.02
5	{BIRADS_breast_density=scattered_fibroglandular_densities, HRT=No, biopsy=Yes } =>{Race_cancer_history=Non-Hispanic-White_Yes}	40	92	1.01

of positive observations that were classified incorrectly, and FP is the number of negative samples that were not classified correctly.

The Area under the Receiver Operating Characteristic curve (ROC) was also considered (Fawcett, 2006), and a detailed description of this metric can be found in Kabir and Ludwig (2019). The F1 measure is another popular performance metric for evaluating classification techniques, which is defined in Equation (6.5)

$$F1\ measure = 2\left(\frac{precision \times recall}{precision + recall}\right) \tag{6.5}$$

The G-mean, which shows the balance between classification performance on the majority and minority classes, was also considered. This metric consists of both positive and negative examples. G-mean can be described as the square root of the product of sensitivity and specificity as shown in Equation (6.6).

$$G\ mean = \sqrt{Sensitivity \times Specificity} \tag{6.6}$$

6.6.2.1 Super Learner Results

We also conducted a comparison of the SL performance to the individual ML algorithms (base learners). We applied the SL techniques to the training data discussed earlier and shown in Table 6.2. To evaluate the model, we applied the test data set as discussed in 6.4.1. Table 6.7 shows the performance (accuracy, precision, recall/sensitivity, specificity) of SL and three different ML methods on the test data.

Table 6.7 Performance (accuracy, precision, recall/sensitivity, specificity) of SL and three ML techniques on the test data (**Bold** indicates the best value for each metric).

Algorithms	Accuracy (%)	Precision (%)	Recall/Sensitivity (%)	Specificity (%)
GBM	88.92	98.14	89.923	73.20
RF	88.88	98.12	89.12	73.02
DNN	**89.50**	97.67	**91.00**	65.95
SL	88.81	**98.17**	89.77	**73.72**

Table 6.8 Performance (AUC, F1, and G-mean) of the SL and three ML algorithms on the test data (**Bold** indicates the best value for each metric).

Algorithms	AUC	F1	G-mean
GBM	0.9234	0.9385	0.8113
RF	0.9198	0.9383	0.8102
DNN	0.7877	**0.9422**	0.7747
SL	**0.9247**	0.9378	**0.8135**

Table 6.8 shows the performance (AUC, F1, and G-mean) of the SL alongside three ML algorithms on the test data.

Comparing Tables 6.7 and 6.8, if we consider the predictive performance based on the test data, we noticed that DNN and SL provide the best results. In Table 6.7, for accuracy and recall, the best values were obtained with DNN. However, the best values for precision and specificity were obtained using the SL method.

In Table 6.8, the performance measures AUC, F1, and G-means were listed. They are considered essential metrics for imbalanced data. In the case of F1, DNN provided the best value, which was slightly higher than the value for SL. However, the best AUC and G-means values were provided by SL, which were slightly greater than for the GBM and RF methods and considerably more significant than for the DNN model.

6.6.3 Discussion

The data mining approach rule discovery, is advantageous as rules can provide meaningful information in a human-readable manner. The class association technique used in Kabir et al. (2018) uncovered knowledge in the form of rules for both breast cancer and non-breast cancer patients from the BCSC risk factor data. From this information, risk factors associated with breast cancer can be learned. However, the data can also be disaggregated by race/ethnicity to generate similar information based on our knowledge of ethnic/racial variations in cancer incidence.

This paper discussed this issue by identifying relevant information in the form of rules for three breast cancer patient race/ethnicity groups.

In ML, supervised learning is used to classify the unknown or target class as accurately as possible for each instance in the data. Various classification algorithms have been applied to breast cancer risk factor data (Kabir and Ludwig, 2018). In this paper, we attempted to enhance the model's performance by applying the super learning model.

The study that we conducted has a few shortcomings. First, the BCSC risk factor data that we investigated for this research is robust, however, we did not know the data's overall quality. Second, there were very few records for breast cancer patients belonging to native American, non-Hispanic Black, or Other/Mixed race/ethnicity groups; these groups had to be removed from the rule mining process. In the rule exploration process, we considered the non-Hispanic white, Hispanic, and Asian/Pacific islander groups. Among these three groups, the number of records for Hispanic and Asian/Pacific islander breast cancer patients was insufficient, compared to records for the non-Hispanic white group.

To address this issue in, we specified multiple support values for the Hispanic and Asian/Pacific islander groups. We also specified a low minimum support due to the low record counts for these two groups. Kabir et al. (2018), Liu et al. (1999) applied the same concept by specifying multiple support values for rare item problems. By applying the same concept, such as setting a low support value, we were able to extract rules for both Hispanic and Asian/Pacific islander groups despite their low number of records in the risk factor data. Although minimum support values were very low for breast cancer patients of the specified race/ethnicity groups, the confidence value that indicates the rules' predictive strength was high. Third, we used resampled training data for a classification model that was obtained using SMOTE and ENN techniques to correct for the highly imbalanced data set – that is, the unequal distributions between cancer and non-cancer individuals. Our SL approach achieved a satisfactory performance.

6.7 Conclusion and Future Work

In this chapter, the class association rule discovery data mining technique and a ML method called SL were studied on breast cancer risk factors. In the first part of this study, rules were extracted from distinct race/ethnicity groups of breast cancer patients, as it is well-known that there are racial/ethnic disparities in cancer prevalence. The experimental results revealed that the produced rules held to the highest confidence level. We also interpreted some crucial rules, showing that they can be easily understood. Physicians or primary care providers can improve their decision-making by analyzing these rules. Targeted prevention plans or processes are also possible for the primary stage of disease or cancer progression. This research can be improved by discovering additional rules for breast cancer patients belonging to other race/ethnicity groups and for patients with no breast cancer according to their race/ethnicity.

Classification is a significant ML task for correctly classifying the target class for each instance in the data. The second part of this discussion concentrated on enhancing the classification model's performance with the SL technique. For a classification task, we applied resampling training data that was obtained using SMOTE and ENN techniques. These resampling methods were performed as the data set used for this study was highly imbalanced, meaning that there was an unequal distribution between cancer and non-cancer patients. The results showed that applying SL to the risk factor data provided a satisfactory predictive performance compared to the individual ML algorithms that were employed as the base learners for this study. To further improve the SL technique's performance, appropriate resampling techniques for this particular data set must be found or developed. This work can also be extended with more diverse methods and optimal parameters to enhance its accuracy. Moreover, as SL usually provides better performance than the individual learners, the technique can be applied to other research problems.

References

A.J. Agboola, A.A. Musa, N. Wanangwa, T. Abdel-Fatah, C.C. Nolan, and et al. Molecular characteristics and prognostic features of breast cancer in nigerian compared with uk women. *Breast Cancer Research and Treatment*, 135(2): 555–569, 2012.

C.C. Aggarwal and P.S. Yu. Mining large itemsets for association rules. *IEEE Data Engineering Bulletin*, 21(1):23–31, 1998.

R. Agrawal, S. Ghosh, T. Imielinski, B. Iyer, and A. Swami. An interval classi er for database mining applications. In *Proceedings of the VLDB Conference*, pages 560–573. Citeseer, 1992.

R. Agrawal, T. Imieliński, and A. Swami. Mining association rules between sets of items in large databases. In *Proceedings of the 1993 ACM SIGMOD international conference on Management of Data*, pages 207–216, 1993.

S. Aiello, C. Click, H. Roark, L. Rehak, and P. Stetsenko. Machine learning with python and h2o. *Edited by Lanford, J., Published by H*, 20:2016, 2016.

W.E. Barlow, E. White, R. Ballard-Barbash, P.M. Vacek, L. Titus-Ernstoff, and et al. Prospective breast cancer risk prediction model for women undergoing screening mammography. *Journal of the National Cancer Institute*, 98(17): 1204–1214, 2006.

C.A. Clarke, T.H.M. Keegan, J. Yang, D.J. Press, A.W. Kurian, and et al. Age-specific incidence of breast cancer subtypes: understanding the black–white crossover. *Journal of the National Cancer Institute*, 104(14): 1094–1101, 2012.

T. Fawcett. An introduction to roc analysis. *Pattern recognition letters*, 27(8): 861–874, 2006.

J. Ferlay, I. Soerjomataram, R. Dikshit, S. Eser, C. Mathers, and et al. Cancer incidence and mortality worldwide: sources, methods and major patterns in globocan 2012. *International Journal of Cancer*, 136(5):E359–E386, 2015.

M.H. Gail, L.A. Brinton, D.P. Byar, D.K. Corle, S.B. Green, and et al. Projecting individualized probabilities of developing breast cancer for white females who are being examined annually. *JNCI: Journal of the National Cancer Institute*, 81(24):1879–1886, 1989.

E. Gauthier, L. Brisson, P. Lenca, and S. Ragusa. Breast cancer risk score: a data mining approach to improve readability. In *The International Conference on Data Mining*, pages 15–21. CSREA Press, 2011.

C. Gibbons, S. Richards, J.M. Valderas, and J. Campbell. Supervised machine learning algorithms can classify open-text feedback of doctor performance with human-level accuracy. *Journal of Medical Internet Research*, 19(3):e65, 2017.

S. Gupta, J. Shah, and B.A. Balasubramanian. Strategies for reducing colorectal cancer among blacks. *Archives of Internal Medicine*, 172(2):182–184, 2012.

J. Han, M. Kamber, and et al. Data mining concept and technology. In *The Annual International Symposium on Supply Chain Management*, volume 132, pages 70–72, 2001.

H. Heyn, S. Moran, I. Hernando-Herraez, S. Sayols, A. Gomez, and et al. Dna methylation contributes to natural human variation. *Genome Research*, 23 (9):1363–1372, 2013.

N. Hou, S. Hong, W. Wang, O.I. Olopade, J.J. Dignam, and D. Huo. Hormone replacement therapy and breast cancer: heterogeneous risks by race, weight, and breast density. *Journal of the National Cancer Institute*, 105(18): 1365–1372, 2013.

F. Kabir and S.A. Ludwig. Classification of breast cancer risk factors using several resampling approaches. In *2018 17th IEEE International Conference on Machine Learning and Applications (ICMLA)*, pages 1243–1248. IEEE, 2018.

F. Kabir and S.A. Ludwig. Enhancing the performance of classification using super learning. *Data-Enabled Discovery and Applications*, 3(1):5, 2019.

F. Kabir, S.A. Ludwig, and A.S. Abdullah. Rule discovery from breast cancer risk factors using association rule mining. In *2018 IEEE International Conference on Big Data (Big Data)*, pages 2433–2441. IEEE, 2018.

H. Kaur and S. Batra. Hpcc: An ensembled framework for the prediction of the onset of diabetes. In *2017 4th International Conference on Signal Processing, Computing and Control (ISPCC)*, pages 216–222. IEEE, 2017.

M. Khalilian and S.T. Tabibi. Breast mass association rules extraction to detect cancerous masses. In *2015 International Congress on Technology, Communication and Knowledge (ICTCK)*, pages 337–341. IEEE, 2015.

M.J. Laan. van der, eric c. polley, and alan e. hubbard. 2007."super learner.". *Statistical Applications in Genetics and Molecular Biology*, 6.

E. LeDell. Scalable super learning. *Handbook of Big Data*, 339, 2016.

W. Li, J. Han, and J. Pei. Cmar: Accurate and efficient classification based on multiple class-association rules. In *Proceedings of the 2001 IEEE International Conference on Data Mining*, pages 369–376. IEEE, 2001.

B. Liu, W. Hsu, and Y. Ma. Mining association rules with multiple minimum supports. In *Proceedings of the 5th ACM SIGKDD International Conference on Knowledge Discovery and Data Mining*, pages 337–341, 1999.

A. More. Survey of resampling techniques for improving classification performance in unbalanced datasets, 2016.

T. Nykodym, T. Kraljevic, A. Wang, and W. Wong. Generalized linear modeling with h2o, 2016. URL https://docs.h2o.ai/h2o/latest-stable/h2o-docs/booklets/GLMBooklet.pdf. (accessed on day month Year).

C. Ordonez, C.A. Santana, and L. De Braal. Discovering interesting association rules in medical data. In *ACM SIGMOD workshop on research issues in data mining and knowledge discovery*, pages 78–85. Citeseer, 2000.

R. Paul, T. Groza, J. Hunter, and A. Zankl. Inferring characteristic phenotypes via class association rule mining in the bone dysplasia domain. *Journal of Biomedical Informatics*, 48:73–83, 2014a.

R. Paul, T. Groza, J. Hunter, and A. Zankl. Inferring characteristic phenotypes via class association rule mining in the bone dysplasia domain. *Journal of Biomedical Informatics*, 48:73–83, 2014b.

S.M.M. Rahman, F. Kabir, and F.A. Siddiky. Rules mining from multi-layered neural networks. *International Journal of Computational Systems Engineering*, 1(1):13–24, 2012.

S.M.M. Rahman, F. Kabir, and M.M. Rahman. Integrated data mining and business intelligence. In *Encyclopedia of Business Analytics and Optimization*, pages 1234–1253. IGI Global, 2014.

T. Silwattananusarn, W. Kanarkard, and K. Tuamsuk. Enhanced classification accuracy for cardiotocogram data with ensemble feature selection and classifier ensemble. *Journal of Computer and Communications*, 4(4):20–35, 2016.

S. Stilou, P.D. Bamidis, N. Maglaveras, and C. Pappas. Mining association rules from clinical databases: an intelligent diagnostic process in healthcare. *Studies in Health Technology and Informatics*, (2):1399–1403, 2001.

M.J. Van der Laan and S. Rose. *Targeted learning: causal inference for observational and experimental data*. Springer Science & Business Media, 2011.

J. Vanerio and P. Casas. Ensemble-learning approaches for network security and anomaly detection. In *Proceedings of the Workshop on Big Data Analytics and Machine Learning for Data Communication Networks*, pages 1–6, 2017.

S. Yao, K. Graham, J. Shen, L.E.S. Campbell, P. Singh, and et al. Genetic variants in micrornas and breast cancer risk in african american and european american women. *Breast cancer research and treatment*, 141(3):447–459, 2013.

7

Neuro-Rough Hybridization for Recognition of Virus Particles from TEM Images[1]

*Debamita Kumar and Pradipta Maji**

Biomedical Imaging and Bioinformatics Lab, Machine Intelligence Unit, Indian Statistical Institute, Kolkata, India

7.1 Introduction

Negative stain transmission electron microscopy (TEM) has long been considered for the detection and description of virus particles. The development of TEM not only facilitates the visualization process of virus structures, but also allows extensive examination of viruses, which makes it an indispensable method for virus recognition. One of the main benefits of using TEM for viral diagnosis is that it does not require organism-specific reagents for identifying the pathogenic agent (Goldsmith and Miller, 2009). In microscopy, negative staining is usually practised to contrast the background of TEM images with an optically opaque fluid for better imagery. However, the level of proficiency, maintenance costs, and time involved in manual investigation and development of automated image acquisition process entail the study of TEM images for virus recognition.

The morphological properties of virus particles, obtained from the TEM images, can be utilized to group particles into different virus classes. For example, Astro-, Rota-, and Adenoviruses exhibit an icosahedral structure, viruses like Dengue, Lassa, and Cowpox are regular in shape, whereas Influenza, Ebola, and Marburg viruses reflect significant amount of irregularity in their structures. Early approaches involve fundamental knowledge regarding the shape and size of the particles for virus identification (Almeida, 1963; Almeida and Waterson, 1970). Besides morphological properties, surface textures play a significant role for categorizing viruses into multiple classes. Different surface textures can be observed for different viruses, when imaged through TEM. Various texture analysis techniques have been developed based on this information for automatic identification of viruses from TEM images (Wen et al., 2016).

*Corresponding Author: Pradipta Maji; debamita_r,pmaji@isical.ac.in

[1]This publication is an outcome of the R&D work undertaken in the project under the Visvesvaraya PhD Scheme of Ministry of Electronics and Information Technology, Government of India, being implemented by Digital India Corporation.

Applied Smart Health Care Informatics: A Computational Intelligence Perspective, First Edition.
Edited by Sourav De, Rik Das, Siddhartha Bhattacharyya, and Ujjwal Maulik.

Since the surface of the virus TEM images contains much of the discriminative information, proper representation of surface textures of the images needs to be obtained, so that the intrinsic properties of each of the virus classes can be highlighted. There are several texture descriptors that encapsulate different textural properties of the images. Certain virus classes can be efficiently represented by some texture descriptors, whereas other descriptors may describe different virus classes accurately. Hence, proper identification of the descriptors is essential for efficient representation of the virus images. However, the inter-class structural similarities, intra-class variations, overlapping class characteristics, and presence of noise enhance the difficulty of virus identification. So, it is required to transform the texture feature representation of the virus images into an output feature space where the images of different classes can be discriminated accurately. Although the discriminative features are automatically learned in deep architectures, but training of the models is computationally very expensive, and also the performance of the architecture depends on the size of training data, parameter optimization, and so on. With a limited training data set, the parameters can not be efficiently learned to reflect the hidden distribution of the given input space.

In this regard, a new architecture is presented in the chapter for automatic identification of the virus particles from negative TEM images. The proposed method judiciously integrates the merits of local texture descriptors, rough hypercuboid approach of rough sets, and restricted Boltzmann machine (RBM). Since local texture descriptors encode intensity variations of pixels within a small neighborhood, different local descriptors are considered in the proposed method to extract textural properties of the TEM images. In order to obtain an efficient representation for each of the virus classes, important features are identified from the class-pair relevant descriptors. The concept of rough hypercuboid approach (Maji, 2014) is used to evaluate the relevance of a descriptor in classifying samples from a particular pair of classes. The theory of rough sets (Pawlak, 1991) is an effective paradigm that provides a mathematical framework to deal with the uncertainties associated with incompleteness in class definitions. The discriminative features are then learned from the input class-pair specific representation of the virus images using the contrastive divergence algorithm of discriminative RBM. Finally, support vector machine (SVM) with linear kernel is employed to classify the latent representation of the TEM images into multiple virus classes. The effectiveness of the proposed architecture, along with a comparison with state-of-the-art methods, is demonstrated on a benchmark database.

7.2 Existing Approaches for Virus Particle Classification

Different surface textures are exhibited by different virus particles when imaged using TEM. So far, various texture analysis techniques have been developed for the automatic identification of virus particles from TEM images. In order to distinguish the four icosahedral structured viruses, namely, Astro, Adeno, Calici, and Rota, an automated method has been proposed in Matuszewski and Shark (2001), that transforms the spatial information of the TEM images into frequency domain with the help of discrete Fourier transform, and thereafter extracts efficient features from the several spectral rings from the magnitude spectrum of images. In Ong (2006), the texture and contour information have been captured from the same four icosahedral virus images using higher order spectral features for propoer identification of the virus structures. Sintorn et al. (2004) have developed a circular template matching method based on radial density profile (RDP) for describing the three maturation stages of human cytomegalovirus capsids present in TEM images. In Kylberg et al. (2011), a concept of RDP, termed as FRDP, has also been used for virus identification, but here instead of spatial domain, it has been computed on Fourier domain. Moreover, Harandi et al. (2014) have introduced an approach to compute and compare covariance descriptors (CovDs) in infinite-dimensional spaces, which are derived by mapping the original data to Hilbert space. The differences between the obtained CovDs in Hilbert space are captured by several Bregman divergences, out of which Jeffreys and Stein divergences based on the SVM have the achieved best performance. The concept of local binary pattern (LBP) (Ojala et al., 1996) has been incorporated with that of RDP in Kylberg et al. (2011) to recognize various virus structures from TEM images based on random forest classifier. Two methods, namely, Fusion and NewH, have been presented by Nanni et al. (2013). While the former method is an ensemble of local phase quantization variants with ternary encoding, the latter concentrates to extract descriptors from the co-occurrence matrix with the objective of improving the efficacy of Haralick's descriptors (Haralick et al., 1973). In Ito et al. (2018), a new architecture has been developed that uses a convolutional neural network to simultaneously learn both features and classifier for indicating where virus particles exist in the TEM image.

In recent years, there has been a growing interest in the field of discriminative learning based approaches for the classification of virus TEM images. Wen et al. (2016) proposed a three-stage architecture to recognize virus particles from negative TEM images. At first, multi-scale principal component analysis filters (MP) are learned from the virus images, then multi-scale completed LBP (MC) features are extracted from the filtered images. Finally, SVM is adopted to classify the images based on the concatenated features, denoted as MPMC. Combining information divergence and dictionary learning (IDDL), a method has been proposed by Cherian et al. (2017) for virus particle identification. It not only learns application specific measures on symmetric positive definite matrices automatically, but also embeds them as vectors using a learned dictionary. A novel distance metric learning algorithm, denoted as threshold auto-tuning metric learning (TATML), has been developed by Onuma et al. (2018), based on the Dykstra algorithm for determining the parameters of distance function. In KGW-SVM (Zhang et al., 2019) method, the distance between two images is measured in terms of the kernel Gauss-Wasserstein (KGW) distance between the corresponding feature representations, which can be low-level features or learned features extracted from deep convolutional neural networks. Approx LogHS and Qapprox LogHS (Minh et al., 2016) provide random Fourier and Quasi-random Fourier approximation methods to formulate the Log-Hilbert-Schmidt distance between covariance operators corresponding to both handcrafted as well as convolutional features, which are efficiently used to classify virus images into multiple classes. Faraki et al. (2015) introduced a novel covariance discriminant learning (CDL) method to provide finite-dimensional approximations of the infinite-dimensional region covariance descriptors for virus image classification by exploiting two feature mappings, namely, random Fourier features and the Nyström method. The heterogeneous auto-similarities of characteristics (HASC) method, introduced by San Biagio et al. (2013), has been applied to heterogeneous dense features maps, encoding linear relations by covariances and non-linear associations through information measures such as mutual information and entropy. Wang et al. (2012) proposed a novel discriminative learning approach, termed as COV, has been proposed where the problem of virus image classification is formulated as the problem of classifying points lying on a Riemannian manifold spanned by the non-singular covariance matrices.

7.3 Proposed Algorithm

In this section, the proposed architecture for the identification of virus particles from negative TEM images is described. The block diagram of the proposed algorithm is demonstrated in Figure 7.1. From Figure 7.1, it can be seen that the proposed method consists of four different stages, namely,

1. Extraction of local textural features from TEM images;
2. Selection of class-pair relevant features using rough hypercuboid approach;
3. Extraction of discriminative features using RBM; and
4. Classification of virus particles using SVM.

In the following subsections, significance of each of the four steps is discussed in details.

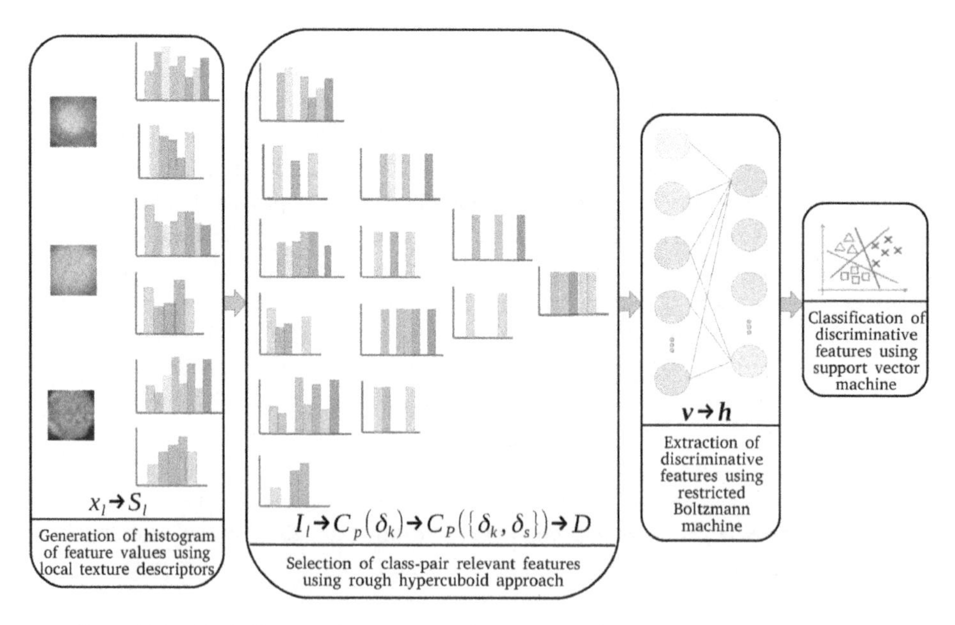

Figure 7.1 Block diagram of the proposed method for the automatic identification of virus particles from TEM images.

7.3.1 Extraction of Local Textural Features

Several texture descriptors are generally considered to characterize the inherent textural properties of the virus TEM images. These descriptors are applied on the input images to extract features that signify the surface texture of the image. However, there are various descriptors that encapsulate different textural properties of the images. Certain textural features may found to be useful in describing the significant properties of a particular virus class, while another set of features may be efficient in depicting the important characteristics of a different virus class. So, in the current study, each virus image is described in terms of the feature values corresponding to several sets of local descriptors, computed at a particular scale.

Let, $\mathbb{U} = \{x_1, \cdots, x_l, \cdots, x_n\}$ be the training set of n virus TEM images, where $x_l \in \Re^m$, and $\mathbb{C} = \{F_1, \cdots, F_i, \cdots, F_m\}$ represents a set of m features. It is assumed that each x_l of $\mathbb{U}$ corresponds to one of the c virus classes $\mathbb{U}/\mathbb{D} = \{\delta_1, \cdots, \delta_k, \cdots, \delta_c\}$, where the set $\mathbb{D}$ contains the image category information. Now, each of the virus TEM images $x_l \in \mathbb{U}$ can be characterized by a set of m feature values $S_l = \{S_{l1}, \cdots, S_{li}, \cdots, S_{lm}\}$, where $S_{li} = x_l(F_i)$ denotes the value of the i-th feature F_i for l-th image x_l. In the current study, four local texture descriptors, namely, LBP: local binary pattern (Ojala et al., 1996); LBP^{ri}: rotation-invariant LBP (Ojala et al., 2002); LBP^{riu2}: rotation-invariant uniform LBP (Ojala et al., 2002); and CoALBP: co-occurrence of adjacent LBP (Nosaka et al., 2012) are considered at four different scales: S_1, S_2, S_3 and S_4. So, S_l represents the normalized histogram for the image x_l obtained from either of the four local descriptors, obtained at a given scale.

Given the input feature set, containing all the features corresponding to each of the descriptors, a feature selection algorithm is introduced in the next step for the identification of class-pair relevant features.

7.3.2 Selection of Class-Pair Relevant Features

The objective of this step is to obtain a compact representation of the input virus images, in terms of the textural features extracted from the images. In general, the number of training samples is limited, while the cardinality of the input feature set is much larger as compared to the number of input samples. So, appropriate learning of the RBM in third stage may become difficult, which may result into inaccurate non-linear feature representation of the virus particles. Hence, it is required to identify relevant features from the input feature space that can properly define each of the virus classes. The proposed method is based on the assumption that a particular descriptor at a given scale may be relevant in classifying a specific pair of virus classes, but may not be able to encapsulate the inherent characteristics of another pair of classes. Therefore, the class-pair relevant feature selection method is proposed in the current study. It first identifies important features from the relevant texture descriptors, selected for each of the virus classes and then obtains the final feature set corresponding to all the classes.

Generally, the fundamental properties of x_l are described in terms of the feature values of the corresponding normalized histogram S_l. However, in the proposed method, it is assumed that all the feature values of S_l do not participate uniformly in illustrating the properties of the TEM image x_l. Indeed, important properties of x_l can be efficiently described with only a subset of features of $\mathbb{C}$, which is referred to as the important feature set I_l of x_l, where $I_l \subseteq \mathbb{C}$. Since the inherent properties of each of the virus images is different from one another, the important feature set is considered to be image specific. Let S_l be sorted in descending order and represented by $\tilde{S}_l = \{\tilde{S}_{l1}, \cdots, \tilde{S}_{li}, \cdots, \tilde{S}_{lm}\}$ such that $\tilde{S}_{l1} \geq \tilde{S}_{l2} \geq \cdots \geq \tilde{S}_{lm}$, and the corresponding feature indices of $\tilde{S}_l$ are preserved in the set $J_l = \{J_{l1}, \cdots, J_{li}, \cdots, J_{lm}\}$.

In order to identify the important features of x_l, the cumulative sum $\psi(x_l, f)$ of the first f features of the sorted normalized histogram $\tilde{S}_l$ is determined, which is termed as the energy function in the proposed algorithm. It denotes the faction of total energy, contained in $\tilde{S}_l$, which is preserved by its' first f features. Hence, important information about x_l can be efficiently described in terms of the energy of x_l computed from $\tilde{S}_l$. The average number of important features $\overline{\mathrm{d}}$ is computed from the individual d_ls, which corresponding to the entire set $\mathbb{U}$. The first $\overline{\mathrm{d}}$ features of the sorted histogram $\tilde{S}_l$ forms the important feature set I_l of the sample x_l, as defined by Liao et al. (2009), which can be expressed as follows:

$$I_l = \{F_i | J_{lf} = i \text{ and } f \leq \overline{\mathrm{d}}\}. \tag{7.1}$$

Thus, the inherent properties of the image x_l can be efficiently described by the features present in the important feature set $I_l \subseteq \mathbb{C}$ of x_l.

Now, the samples of same class are expected to have similar important features, while the samples of different classes are assumed to have different important feature sets. Hence, the probability of occurrence $P(F_i|\delta_k)$ of feature F_i in the important feature sets of samples corresponding to a specific class δ_k is computed at first, and then, noisy features are identified and discarded by applying a threshold ϵ on the set. The features F_i having $P(F_i|\delta_k) \geq \epsilon$ constitute a set $C(\delta_k)$, which signifies important characteristics of the class δ_k. So, the features F_i present in $C(\delta_k)$ not only characterize the intrinsic properties of images belonging to the class δ_k but also illustrate the inherent characteristics of the class δ_k. Let us assume that the set $C(\{\delta_k, \delta_s\})$ represents the significant properties of the class-pair $\{\delta_k, \delta_s\}$. Then, it should include features that are able to depict the characteristics of both the virus classes δ_k and δ_s. So, $C(\{\delta_k, \delta_s\})$ contains only those features F_i which are present in both the sets $C(\delta_k)$ and $C(\delta_s)$.

Let $\mathcal{M} = \{\mathcal{M}_1, \cdots, \mathcal{M}_p, \cdots, \mathcal{M}_t\}$ be the set of modalities considered. In the proposed method, a modality denotes a particular local texture descriptor computed at a specific scale. The relevance $\Gamma_p(\{\delta_k, \delta_s\})$ of the set $C_p(\{\delta_k, \delta_s\})$, corresponding to the modality $\mathcal{M}_p \in \mathcal{M}$, is defined to quantify the performance of the feature set in demonstrating the inherent characteristics of the pair of classes $\{\delta_k, \delta_s\}$.

The concept of a hypercuboid equivalence partition matrix of the rough hypercuboid approach (Maji, 2014) is employed to quantify the relevance of a feature set. The relevance $\Gamma_p(\{\delta_k, \delta_s\})$ of the feature set $C_p(\{\delta_k, \delta_s\})$ with reference to the class-pair $\{\delta_k, \delta_s\}$ is obtained as follows (Maji, 2014):

$$\Gamma_p(\{\delta_k, \delta_s\}) = 1 - \frac{1}{n_{ks}} \sum_{l=1}^{n_{ks}} v_l(C_p(\{\delta_k, \delta_s\})) \tag{7.2}$$

where n_{ks} is the number of training images corresponding to the pair of virus classes $\{\delta_k, \delta_s\}$ and

$$\nu_l(C_p(\{\delta_k, \delta_s\})) = \begin{cases} 1 & \text{if } \mu_{kl}(C_p) = 1 \text{ and } \mu_{sl}(C_p) = 1 \\ 0 & \text{otherwise,} \end{cases} \tag{7.3}$$

$$\text{where } \mu_k(C_p) = [\mu_{kl}(C_p)]_{1\times n_{ks}} = \bigcap_{F_i \in C_p} \mu_k(F_i). \tag{7.4}$$

Here, $\mu_{kl}(F_i) \in \{0, 1\}$ signifies the belongingness of sample x_l in the k-th rough hypercuboid equivalence partition formed by the i-th feature F_i and is obtained as follows:

$$\mu_{kl}(F_i) = \begin{cases} 1 & \text{if } \mathrm{L}(\delta_k) \leq x_l(F_i) \leq \mathrm{U}(\delta_k) \\ 0 & \text{otherwise.} \end{cases} \tag{7.5}$$

The interval $[\mathrm{L}(\delta_k), \mathrm{U}(\delta_k)]$ corresponds to the range of values that the feature F_i attains for samples belonging to the class δ_k. Clearly, $\forall F_i \in C_p(\{\delta_k, \delta_s\}), x_l(F_i) \in [\mathrm{L}(\delta_k), \mathrm{U}(\delta_k)]$ if $x_l \in \delta_k$. In other words, the k-th equivalence partition induced by the feature set $C_p(\{\delta_k, \delta_s\})$ should atleast contain the samples belonging to the class δ_k, which makes an equivalence partition non-empty. An implicit hypercuboid, depicted by the shaded rectangle in Figure 7.2, is formed at the overlapping region of hypercuboids, induced by the features of the set $C_p(\{\delta_k, \delta_s\})$. It encompasses all the samples that are misclassified by the set $C_p(\{\delta_k, \delta_s\})$, where each of the samples is present in more than one equivalence partition. Based on the cardinality of the implicit hypercuboid, the relevance $\Gamma_p(\{\delta_k, \delta_s\})$ of $C_p(\{\delta_k, \delta_s\})$, with reference to the class-pair $\{\delta_k, \delta_s\}$ is determined.

So, it is evident from Equation 7.2 that as the number of samples in the implicit hypercuboid increases, the relevance decreases. If $\Gamma_p(\{\delta_k, \delta_s\}) = 1$, then no overlapping region exists between equivalence partitions. In other words, the classes δ_k and δ_s can be precisely described in terms of the set $C_p(\{\delta_k, \delta_s\})$. If $\Gamma_p(\{\delta_k, \delta_s\}) = 0$, then the classes cannot be described using the knowledge of $C_p(\{\delta_k, \delta_s\})$. However, if

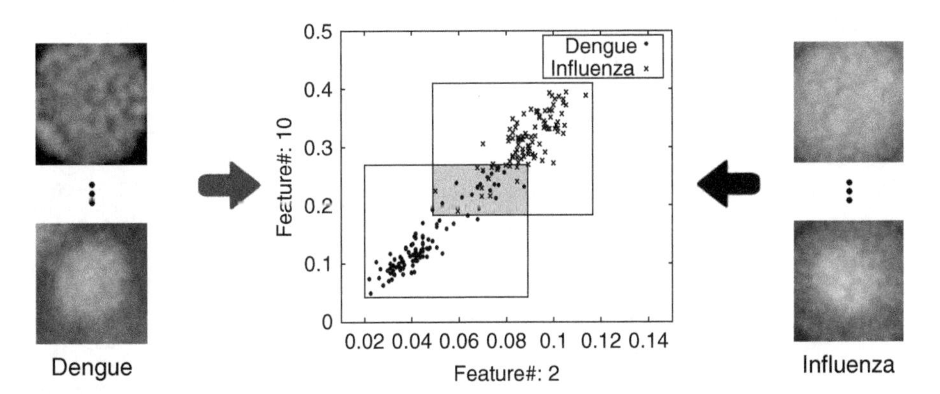

Figure 7.2 An illustration of two class rough hypercuboids, considering the upper approximations of Dengue and Influenza virus classes, presented in two dimensions. The implicit hypercuboid is depicted by the shaded region.

$\Gamma_p(\{\delta_k, \delta_s\}) \in (0, 1)$, then δ_k and δ_s can be approximated with reference to the information of the feature set $C_p(\{\delta_k, \delta_s\})$. Based on the relevance of $C_p(\{\delta_k, \delta_s\})$, which considers each of the t modalities, the class-pair relevant feature set $\tilde{C}_{ks}$ that corresponds to the pair of classes $\{\delta_k, \delta_s\}$ is formed, which is given by

$$\tilde{C}_{ks} = \arg\max_{C_p(\{\delta_k, \delta_s\})}\{\Gamma_p(\{\delta_k, \delta_s\})\}. \tag{7.6}$$

Eventually, the final feature set $\mathcal{D}$ is achieved as the union of class-pair relevant feature sets $\tilde{C}_{ks}$ for all given pair of virus classes:

$$\mathcal{D} = \bigcup \tilde{C}_{ks}. \tag{7.7}$$

So, the feature set $\mathcal{D}$ provides an effective representation of the virus particles, corresponding to all the classes. This reduced set of features is then applied to the visible units of discriminative RBM to learn non-linear discriminative features for each of the training virus images.

7.3.3 Extraction of Discriminating Features

In this step, the latent distribution of the training virus samples is learned with the help of an RBM, where the samples are described using the feature set $\mathcal{D}$ of (7.7). The basic objective of training an RBM is to apply a non-linear transformation to the input feature space such that the samples belonging to different classes are well separated in the mapped feature space, leading to an accurate classification of the virus samples. Primarily, RBM is an unsupervised network, but the proposed method avails the class label information for virus identification. So, if the class label information can be incorporated into the extracted features, then the discriminative performance of the model can be improved. Hence, in the proposed method, discriminative RBM (Larochelle and Bengio, 2008) is employed, where the observed variables are partitioned between visible units and class units.

Let the cardinality of input feature set be r, that is, $|\mathcal{D}| = r$. So, the RBM in the current study consists of r visible units $\boldsymbol{V} = (V_1, \cdots, V_g, \cdots, V_r)$, c class units $\boldsymbol{Y} = (Y_1, \cdots, Y_k, \cdots, Y_c)$ and q hidden units $\boldsymbol{H} = (H_1, \cdots, H_j, \cdots, H_q)$, where q is assumed to be the dimension of target distribution. The random variables $(\boldsymbol{V}, \boldsymbol{Y}, \boldsymbol{H})$ can take any value $(\boldsymbol{v}, \boldsymbol{y}, \boldsymbol{h}) \in \{0, 1\}^{r+c+q}$ which define the state of the model at that instance. The visible units are initialized with the input feature values corresponding to each virus images. Therefore, for a training sample $x_l \in \mathbb{U}, v_g = x_l(F_i), \forall F_i \in \mathcal{D}$, where $g \in \{1, 2, \cdots r\}$ and $y_k = 1$ if $x_l \in \delta_k$, otherwise the state of all other class units remains at zero. The energy function of the model is given by

$$E(\boldsymbol{v}, \boldsymbol{y}, \boldsymbol{h}) = -\sum_{g=1}^{r}\sum_{j=1}^{q} v_g w_{gj} h_j - \sum_{k=1}^{c}\sum_{j=1}^{q} y_k u_{kj} h_j - \sum_{g=1}^{r} a_g v_g - \sum_{j=1}^{q} b_j h_j - \sum_{k=1}^{c} o_k y_k. \tag{7.8}$$

The bidirectional weight parameters w_{gj} and u_{kj} connect the visible unit V_g and hidden unit H_j, and class unit Y_k and H_j, respectively. The bias parameters a_g, b_j, and o_k

are associated with V_g, H_j, and Y_k, respectively. Given the states of visible and class units, the probability of hidden units being on is given by

$$p(h_j = 1|\boldsymbol{v},\boldsymbol{y}) = \sigma\left(\sum_{g=1}^{r} v_g w_{gj} + \sum_{k=1}^{c} y_k u_{kj} + b_j\right); \tag{7.9}$$

$$\text{where } \sigma(z) = \frac{1}{1+\exp(-z)} \tag{7.10}$$

is the sigmoid function. So, the hidden representation of the model is obtained by applying a non-linear transformation to the input units. Hence, upon proper learning, the RBM can be considered a non-linear feature detector.

Now, learning of the RBM refers to updating the parameters of the model in such a way that the latent distribution fits the given observations as much as possible. In the proposed method, contrasting divergence (Hinton, 2002) learning, based on gradient ascent, is performed to estimate the given parameters of the model. Once the parameters are properly learned, given the input feature set $\mathcal{D}$, the output feature representation $\mathcal{H}_l = \{\mathcal{H}_1, \cdots, \mathcal{H}_j, \cdots, \mathcal{H}_q\}$ for the virus sample x_l can be obtained using (7.9), where $\mathcal{H}_j = p(h_j = 1|v, y), \forall j$. Thus, discriminative features are extracted using the RBM, which are then used to classify the virus images into multiple classes.

7.3.4 Classification

The SVM (Vapnik, 1995), which is based on supervised statistical learning theory, is considered in the proposed method for classification purpose. Given the discriminative feature representation $\mathcal{H}$ corresponding to the training virus samples, the SVM attempts to obtain a hyperplane or decision boundary in the specified feature space that maximizes the difference between the hyperplane and support vect well as minimizes the classification error between the given classes. The expression for q-dimensional hyperplane $\tilde{\mathcal{H}}$ is given by

$$\boldsymbol{\omega} \cdot \tilde{\mathcal{H}} - \beta = 0, \tag{7.11}$$

where $\boldsymbol{\omega}$ is the vector normal to the hyperplane, and the parameter $\frac{\beta}{\|\boldsymbol{\omega}\|}$ denotes the offset of the hyperplane $\tilde{\mathcal{H}}$ from the origin along the normal vector $\boldsymbol{\omega}$. So, the objective of SVM classifier is to minimize the following function:

$$\varphi(\boldsymbol{\omega}, \beta) = \frac{1}{n}\left(\sum_{l=1}^{n} \xi_l + \frac{\lambda}{2}\|\boldsymbol{\omega}\|^2\right), \tag{7.12}$$

subject to

$$\gamma_l(\boldsymbol{\omega} \cdot \mathcal{H}_l) > 1 - \xi_l \text{ and } \xi_l > 0, \forall l.$$

Here, the term λ accounts for the trade-off between margin width and rate of misclassification and $\gamma_l \in \{+1, -1\}$ indicating the class to which the sample x_l belongs, considering a two-class problem. In the case of test samples, the feature set $\mathcal{D}$, obtained in Equation 7.7, is applied to the input of RBM and the corresponding discriminative feature representation $\mathcal{H}$ is obtained in Equation 7.9. Finally, the class label of test virus sample is predicted based on the decision boundary obtained using Equation 7.11.

7.4 Experimental Results and Discussion

The proficiency of the proposed method in recognizing the virus structures from the negative stain TEM images is analyzed extensively on Virus image data set and the corresponding results are reported in this section, along with a comparison with related methods.

7.4.1 Experimental Setup

In order to validate the efficacy of the proposed method, four local texture descriptors, namely, LBP (Ojala et al., 1996), LBP^{ri} (Ojala et al., 2002), LBP^{riu2} (Ojala et al., 2002) and CoALBP (Nosaka et al., 2012), evaluated at scales S_1: scale 1; S_2: scale 2; S_3: scale 3; S_4: scale 4; S_{123}: concatenation of S_1, S_2, and S_3; and S_{124}: concatenation of S_1, S_2, and S_4 are considered. It is to be noted here that the descriptors and the corresponding scales are considered arbitrarily in the current study, and therefore, the proposed descriptor selection method is equally compatible to be applied on any other sets of descriptors or scales. Also, for the computation of CoALBP, 4 neighboring pixels are considered, whereas 8-neighborhood is considered for the rest. In the proposed algorithm, the optimal values of parameters ψ_0 and ϵ for the Virus image database are considered to be 0.95 and 0.30, respectively, as obtained in (Kumar and Maji, 2019). The number of hidden units q of the RBM is fixed at 1000, and one-step contrasting divergence learning is considered to approximate the parameters of the given model. The SVM with linear kernel is employed in the proposed architecture for the identification of virus particles from TEM images.

The 10-fold cross-validation (CV) is employed to analyse the performance of the proposed method as well as related approaches. The comparative performance of the algorithms is studied through box-and-whisker plots, tables of means, medians, standard deviations, and p-values computed through both paired-t and Wilcoxon signed-rank tests, with 95% confidence level. In box-and-whisker plots, the median is represented by the central line of the box; upper and lower boundaries depict the upper and lower quartiles, respectively. Whiskers are drawn from the mean to three standard deviations, so that extreme points can also be included. The outliers are plotted with '+', individually.

7.4.2 Methods Compared

The performance of the proposed method is studied with reference to various deep architectures, which include discriminative deep belief network (disDBN) (Liu et al., 2011), deep Boltzmann machine with SVM as classifier (DBM+SVM) (Salakhutdinov and Hinton, 2009), deep convolutional neural network (deepCNN) (Krizhevsky et al., 2012), and deep convolutional auto-encoder (DCAE) (Geng et al., 2015). Also, several state-of-the-art methods are considered for comparative performance analysis, which are Haralick textural features (Haralick et al., 1973), discriminative features for texture description (DFTD) (Guo et al., 2012), dominant local binary pattern (DLBP) (Liao et al., 2009), Fourier radial density profile (FRDP)

(Kylberg et al., 2011), MPMC (Wen et al., 2016), NewH (Nanni et al., 2013), Fusion (Nanni et al., 2013), NewF (Nanni et al., 2014), LBP+RDP (Kylberg et al., 2011), Jeffreys/Stein (Harandi et al., 2014), CDL (Faraki et al., 2015), HASC (San Biagio et al., 2013), TATML (Onuma et al., 2018), IDDL (Cherian et al., 2017), COV+SVM (Wang et al., 2012), Approx LogHS (Minh et al., 2016), Qapprox LogHS (Minh et al., 2016), and KGW-SVM (Zhang et al., 2019).

7.4.3 Database Considered

The effectiveness of the proposed method as well as existing approaches is evaluated on the real-life Virus data set (Kylberg et al., 2012). The database contains 1500 samples which are imaged through negative stain TEM and the entire set is partitioned into ten different folds. Each of the samples in the set belongs to one of the fifteen different virus classes. Although, different viruses are characterized by different structural properties, the cross-section or diameter remains almost constant within a particular virus class. Generally, the diameter varies from 25 nm to 270 nm based on the morphology of the virus particles. The virus classes, considered in the data set include Adenovirus, Astrovirus, Rotavirus, Norovirus, Dengue, Cowpox, Ebola, Influenza, Lassa, Marburg, Orf, Papilloma, Crimean-Congo Haemorrhagic Fever, Rift Valley, and West Nile.

7.4.4 Effectiveness of Proposed Approach

In order to establish the significance of each of the steps employed in the proposed approach, extensive experimentation is carried out on the Virus image data set and the obtained results are reported in Figure 7.3 and Table 7.1. The performance of the proposed method is compared with that of Method_1, Method_2, and Method_3. Here, Method_1 corresponds to the proposed method where the virus TEM images are classified using the SVM based on the input set of local textural features, extracted from the images. The cardinality of the input feature set is 4280. In Method_2, the proposed class-pair relevant feature selection method is considered, that is, the SVM is used to predict the class label of test virus samples based on the feature set D, formed with only 2327 features using Equation 7.7. Finally, Method_3 refers to the proposed method where the feature selection algorithm is not used and the latent distribution of the virus TEM images is estimated directly from the input feature representation of the images. It means the 4280 dimensional input feature space is transformed into a feature space with only 1000 dimensions using the given model.

It can be noticed from Figure 7.3 and Table 7.1 that the classification accuracy has improved for Method_2 from that of Method_1. It signifies that the proposed descriptor selection method can effectively recognize class-pair relevant features to represent each of the virus class pairs with significantly lesser number of features. In the case of Method_3, the performance deteriorated due to inefficient representation of the virus images, resulting into insufficient learning of the model. However, in the proposed method, the latent distribution of the virus samples is appropriately approximated with only 1000 hidden units of RBM; and therefore, considerable

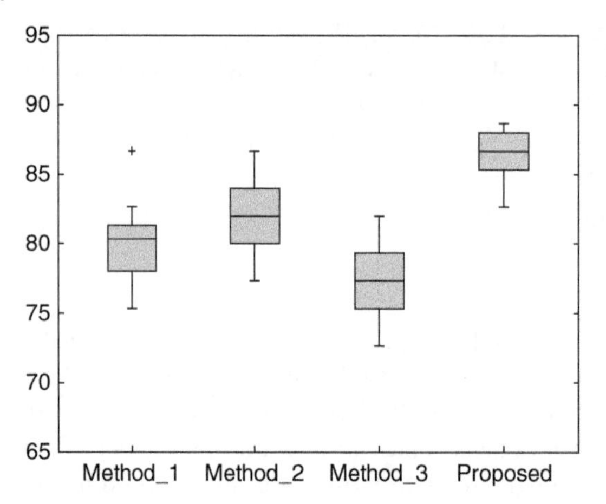

Figure 7.3 Comparative performance analysis of the proposed method and its variants.

Table 7.1 Classification accuracy (%) obtained by the proposed method and its different variants.

Different Folds	Method_1	Method_2	Method_3	Proposed
Fold1	76.00	82.00	75.33	86.67
Fold2	82.67	84.00	72.67	86.67
Fold3	78.00	80.00	76.67	88.00
Fold4	80.67	82.00	75.33	82.67
Fold5	75.33	77.33	79.33	85.33
Fold6	80.00	81.33	82.00	83.33
Fold7	78.67	80.00	78.67	86.00
Fold8	86.67	84.00	77.33	88.67
Fold9	80.67	86.67	79.33	87.33
Fold10	81.33	83.33	77.33	88.67
Mean	80.00	82.07	77.40	**86.33**
StdDev	3.30	2.62	2.62	**2.07**
Median	80.33	82.00	77.33	**86.67**
Wilcoxon:p	2.52E-03	2.52E-03	2.47E-03	–
Paired-*t*:p	9.94E-05	3.78E-04	1.50E-05	–

improvement in the classification accuracy can be noted, irrespective of the folds considered. Also, from the statistical significance analysis, it can be observed that significantly lower p-values are attained by the proposed approach in all the 6 cases, considering 95% confidence interval.

Generally, a particular descriptor, evaluated at a specific scale, is used to capture the inherent characteristics of all the TEM images present in the Virus data set. However, in the proposed approach, the class-pair relevant features are identified

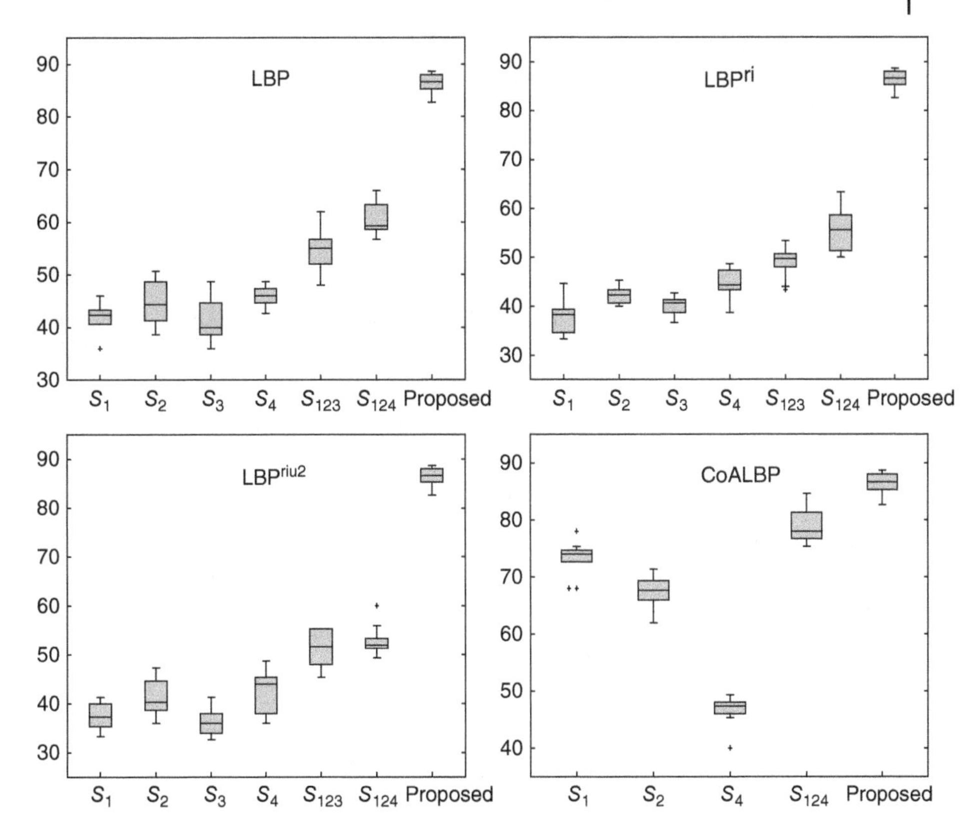

Figure 7.4 Comparative performance analysis between the proposed method and different local texture descriptors.

at first to represent the textural properties of each of the virus classes as well as the intrinsic properties of the TEM images. Thereafter, non-linear transformation is applied to the input feature space and discriminative features are learned using RBM which are then used to classify the virus TEM images into multiple classes. In order to validate the effectiveness of the proposed approach over existing texture analysis approaches, extensive investigation is conducted on the Virus data set, given an input set of 15 modalities corresponding to four local texture descriptors LBP, LBP^{ri}, LBP^{riu2} and CoALBP, evaluated at both the individual and concatenated scales. Figure 7.4 and Table 7.2 depict the classification accuracy achieved by the proposed method as well as various local texture descriptors on the samples of Virus data set. The results presented in Figure 7.4 and Table 7.2 reveal that the proposed algorithm achieves significantly better classification accuracy in all the 44 cases.

7.4.5 Comparative Performance Analysis

This section compares the performance of the proposed method with that of several deep learning methods and existing state-of-the-art virus particle classification approaches.

Table 7.2 Performance analysis of different local descriptors and proposed method.

Descriptors	Scales	Mean	StdDev	Median	Wilcoxon:p	Paired-*t*:p
LBP	S_1	42.00	2.70	42.33	2.53E-003	4.12E-012
	S_2	44.60	4.10	44.33	2.52E-003	1.89E-010
	S_3	41.20	3.91	40.00	2.53E-003	1.61E-011
	S_4	45.73	**1.78**	46.00	2.53E-003	2.52E-013
	S_{123}	55.00	4.21	55.00	2.52E-003	2.93E-009
	S_{124}	60.53	3.20	59.33	2.53E-003	1.10E-008
LBP^{ri}	S_1	37.87	3.45	38.33	2.53E-003	1.05E-011
	S_2	42.33	1.87	42.33	2.50E-003	2.16E-012
	S_3	40.00	2.04	40.67	2.50E-003	2.86E-014
	S_4	44.53	2.96	44.33	2.47E-003	1.83E-011
	S_{123}	48.87	3.19	49.67	2.52E-003	2.95E-010
	S_{124}	55.67	4.36	55.67	2.53E-003	4.97E-009
LBP^{riu2}	S_1	37.53	3.06	37.33	2.53E-003	4.70E-012
	S_2	41.13	3.61	40.33	2.53E-003	8.62E-011
	S_3	36.47	3.19	36.00	2.53E-003	2.80E-012
	S_4	42.40	4.67	44.00	2.53E-003	1.18E-010
	S_{123}	51.20	3.73	51.67	2.52E-003	3.85E-010
	S_{124}	52.93	3.03	52.00	2.52E-003	1.28E-011
CoALBP	S_1	73.20	3.14	74.00	2.52E-003	6.08E-007
	S_2	67.47	2.81	67.67	2.53E-003	2.51E-008
	S_4	46.73	2.64	47.33	2.50E-003	5.05E-012
	S_{124}	78.87	2.92	78.00	2.52E-003	2.03E-005
Proposed		**86.33**	2.07	**86.67**		

7.4.5.1 Comparison with Deep Architectures

In this section, the performance of the proposed approach is compared with that of several existing deep architectures, namely, disDBN (Liu et al., 2011), DBM+SVM (Salakhutdinov and Hinton, 2009), deepCNN (Krizhevsky et al., 2012), and DCAE (Geng et al., 2015). The corresponding results for 10-fold CV are presented in Figure 7.5 and Table 7.3. It is evident from the results that the proposed method performs better than the existing deep architectures in classifying TEM images into different virus classes. It is also found to be statistically significant in all the cases.

This is due to the fact that deep learning models usually require a large training set of samples for identifying latent distribution of the given data. In case of Virus database, the number of training images is limited. Also, poor illumination and lighting problems increase the difficulty of virus particles identification. However,

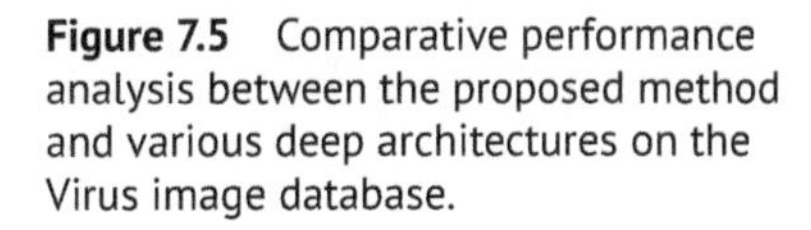

Figure 7.5 Comparative performance analysis between the proposed method and various deep architectures on the Virus image database.

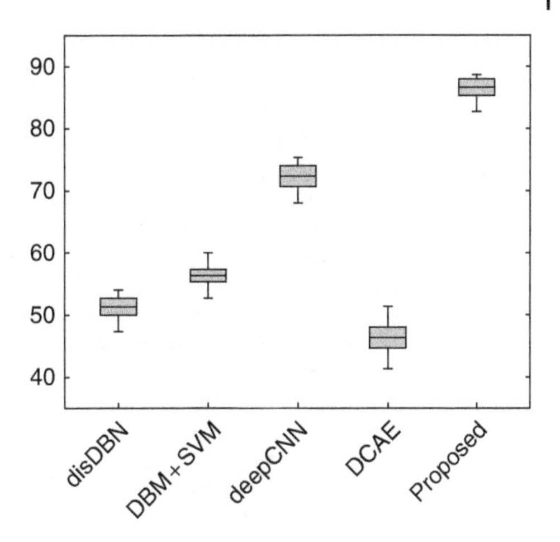

Table 7.3 Performance analysis of the proposed method and several deep architectures.

Algorithms	Mean	StdDev	Median	Wilcoxon:p	Paired-*t*:p
disDBN	51.13	2.25	51.33	2.53E-003	2.28E-010
DBM+SVM	56.40	2.31	56.33	2.46E-003	2.20E-016
deepCNN	72.20	2.18	72.33	2.53E-003	2.66E-007
DCAE	46.27	2.83	46.33	2.52E-003	4.61E-012
Proposed	**86.33**	**2.07**	**86.67**		

in case of proposed approach, efficient representation of the virus images are obtained at first, then latent distribution of the samples is estimated, and thus, considerable performance is achieved on the test set of virus samples.

7.4.5.2 Comparison with Existing Approaches

Finally, the performance of the proposed method is studied with reference to several state-of-the-art methods for virus classification and the corresponding results are reported in Figure 7.6. The existing approaches include various texture analysis methods: MPMC (Wen et al., 2016), NewH (Nanni et al., 2013), Fusion (Nanni et al., 2013), NewF (Nanni et al., 2014), Haralick textural features (Haralick et al., 1973), LBP+RDP (Kylberg et al., 2011), DLBP (Liao et al., 2009), DFTD (Guo et al., 2012), and FRDP (Kylberg et al., 2011); different discriminative learning methods based on covariance descriptors: COV+SVM (Wang et al., 2012), CDL (Faraki et al., 2015), and Jeffreys/Stein (Harandi et al., 2014); and several distance based learning methods: HASC (San Biagio et al., 2013), Approx LogHS (1) and Approx LogHS (2) (Minh et al., 2016), Qapprox LogHS (1) and Qapprox LogHS (2) (Minh et al., 2016), KGW-SVM (1) and KGW-SVM (2) (Zhang et al., 2019), where (1) indicates low-level features and (2) denotes deep features, TATML (Onuma et al., 2018), and IDDL (Cherian et al., 2017).

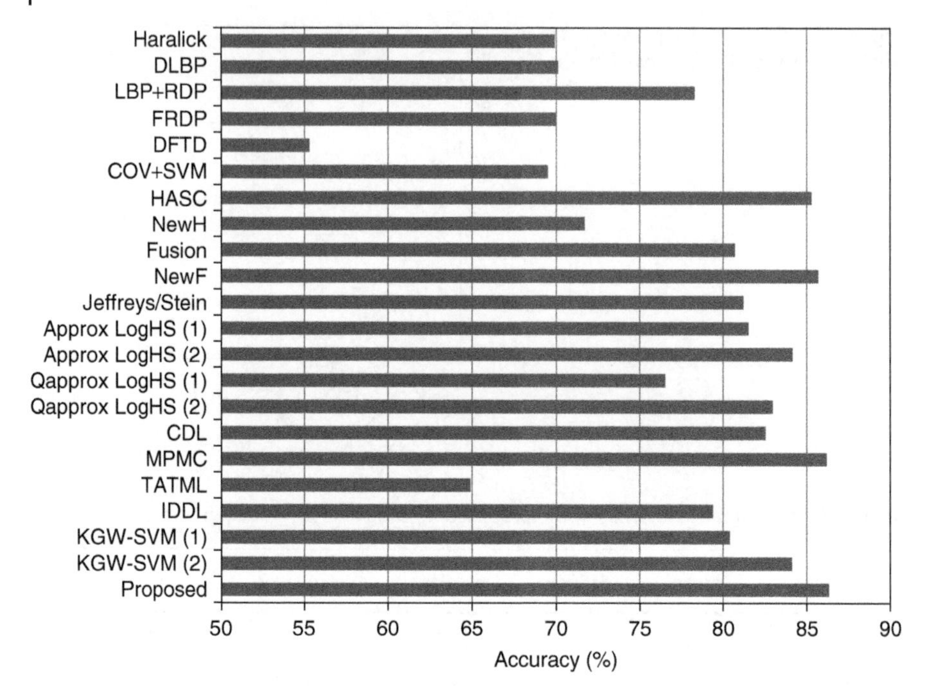

Figure 7.6 Comparative performance analysis of the proposed and existing approaches.

From the results presented in Figure 7.6, it can be seen that the approaches corresponding to the Haralick features, DFTD, DLBP, FRDP, COV+SVM, and TATML achieve poor accuracy in categorizing virus particles from TEM images, whereas the methods like LBP+RDP, Fusion, NewH, Jeffreys/Stein, Approx LogHS (1), Qapprox LogHS (1), Qapprox LogHS (2), IDDL, CDL, and KGW-SVM (1) demonstrate satisfactory results. Although the methods associated with HASC, NewF, MPMC, Approx LogHS (2), and KGW-SVM (2) exhibit improvement in the performance, yet the proposed method achieves highest classification accuracy on the Virus data set with respect to all the existing methods.

7.5 Conclusion

The objective of the current research work is to develop an efficient method for automatic recognition of different virus particles from negative TEM images. In this regard, a novel method has been developed, integrating judiciously the merits of local texture descriptors, rough hypercuboid approach of rough sets, discriminative restricted Boltzmann machine (RBM), and the support vector machine (SVM). In the proposed approach, the intrinsic properties of the TEM images were described by a set of local texture features that are obtained by applying various texture descriptors at different scales. The effective representation of the virus images is then achieved by identifying the important features from the class-pair relevant descriptors corresponding to all the pairs of virus classes. The rough hypercuboid approach helps to identify relevant features for the virus samples belonging to overlapping and incomplete classes. Given the reduced feature set as input, the discriminative features are learned with the help of RBM using contrastive divergence method. Finally, the SVM is adopted to classify the latent representation of the virus images into multiple classes. The efficiency of the proposed method, along with a comparison with several related approaches, has been demonstrated on a real-life Virus image data set.

References

J.D. Almeida. A Classification of Virus Particles Based on Morphology. *Canadian Medical Association Journal*, 89(16):787–798, 1963.

J.D. Almeida and A.P. Waterson. Some Implications of a Morphologically Oriented Classification of Viruses. *Archives of Virology*, 32(1):66–72, 1970.

A. Cherian, P. Stanitsas, M. Harandi, V. Morellas, and N. Papanikolopoulos. Learning Discriminative Alpha-Beta-divergence for Positive Definite Matrices (Extended Version). *arXiv:1708.01741*, 2017.

M. Faraki, M.T. Harandi, and F. Porikli. Approximate Infinite-dimensional Region Covariance Descriptors for Image Classification. In *Proceedings of the IEEE International Conference on Acoustics, Speech and Signal Processing (ICASSP)*, pages 1364–1368. IEEE, 2015.

J. Geng, J. Fan, H. Wang, X. Ma, B. Li, and F. Chen. High-resolution SAR Image Classification via Deep Convolutional Autoencoders. *Geoscience and Remote Sensing Letters*, 12(11):2351–2355, 2015.

C.S. Goldsmith and S.E. Miller. Modern Uses of Electron Microscopy for Detection of Viruses. *Clinical Microbiology Reviews*, 22(4):552–563, 2009.

Y. Guo, G. Zhao, and M. Pietikäinen. Discriminative Features for Texture Description. *Pattern Recognition*, 45(10):3834–3843, 2012.

R. M. Haralick, K. Shanmugam, and I. Dinstein. Textural features for image classification. *IEEE Transactions on Systems, Man, and Cybernetics*, 3(6): 610–621, 1973.

M. Harandi, M. Salzmann, and F. Porikli. Bregman Divergences for Infinite Dimensional Covariance Matrices. In *Proceedings of the IEEE Conference on Computer Vision and Pattern Recognition*, pages 1003–1010, 2014.

G. E. Hinton. Training Products of Experts by Minimizing Contrastive Divergence. *Neural Computation*, 14(8):1771–1800, 2002.

E. Ito, T. Sato, D. Sano, E. Utagawa, and T. Kato. Virus Particle Detection by Convolutional Neural Network in Transmission Electron Microscopy Images. *Food and Environmental Virology*, 10(2):201–208, 2018.

A. Krizhevsky, I. Sutskever, and G.E. Hinton. Imagenet Classification with Deep Convolutional Neural Networks. In *Proceedings of the Advances in Neural Information Processing Systems*, pages 1097–1105, 2012.

D. Kumar and P. Maji. An Efficient Method for Automatic Recognition of Virus Particles in TEM Images. In *Proceedings of the International Conference on Pattern Recognition and Machine Intelligence*, pages 21–31. Springer, 2019.

G. Kylberg, M. Uppström, and I.M. Sintorn. Virus Texture Analysis using Local Binary Patterns and Radial Density Profiles. In *Proceedings of the Iberoamerican Congress on Pattern Recognition*, pages 573–580. Springer, 2011.

G. Kylberg, M. Uppstrom, G. Borgefors, and I.M Sintorn. Segmentation of Virus Particle Candidates in Transmission Electron Microscopy Images. *Journal of Microscopy*, 245:140–147, 2012.

H. Larochelle and Y. Bengio. Classification Using Discriminative Restricted Boltzmann Machines. In *Proceedings of the International Conference on Machine Learning*, pages 536–543, 2008.

S. Liao, M.W.K. Law, and A.C.S. Chung. Dominant local binary patterns for texture classification. *IEEE Transactions on Image Processing*, 18(5): 1107–1118, 2009.

Y. Liu, S. Zhou, and Q. Chen. Discriminative Deep Belief Networks for Visual Data Classification. *Pattern Recognition*, 44(10–11):2287–2296, 2011.

P. Maji. A rough hypercuboid approach for feature selection in approximation spaces. *IEEE Transactions on Knowledge and Data Engineering*, 26(1):16–29, 2014.

B.J. Matuszewski and L.K. Shark. Hierarchical Iterative Bayesian Approach to Automatic Recognition of Biological Viruses in Electron Microscope Images. In *Proceedings of the International Conference on Image Processing*, pages 347–350, 2001.

H.Q. Minh, M.S. Biagio, L. Bazzani, and V. Murino. Kernel Methods on Approximate Infinite-Dimensional Covariance Operators for Image Classification. *arXiv:1609.09251*, 2016.

L. Nanni, M. Paci, S. Brahnam, S. Ghidoni, and E. Menegatti. Virus Image Classification using Different Texture Descriptors. In *Proceedings of the International Conference on Bioinformatics and Computational Biology, Las Vegas, NV*, 2013.

L. Nanni, M. Paci, F.L. Caetano Dos Santos, S. Brahnam, and J. Hyttinen. Analysis of Virus Textures in Transmission Electron Microscopy Images. *Innovation in Medicine and Healthcare*, 207:83–91, 2014.

R. Nosaka, Y. Ohkawa, and K. Fukui. Feature extraction based on co-occurrence of adjacent local binary patterns. In *Proceedings in Advances in Image and Video Technology*, pages 82–91, 2012.

T. Ojala, M. Pietikäinen, and D. Harwood. A comparative study of texture measures with classification based on feature distributions. *Pattern Recognition*, 29(1):51–59, 1996.

T. Ojala, M. Pietikainen, and T. Maenpaa. Multiresolution gray-scale and rotation invariant texture classification with local binary patterns. *IEEE Transactions on Pattern Analysis and Machine Intelligence*, 24(7):971–987, 2002.

H.C.L. Ong. *Virus Recognition in Electron Microscope Images using Higher Order Spectral Features*. PhD thesis, Queensland University of Technology, 2006.

Y. Onuma, R. Rivero, and T. Kato. Threshold Auto-Tuning Metric Learning. *arXiv:1801.02125*, 2018.

Z. Pawlak. *Rough Sets: Theoretical Aspects of Reasoning about Data*. Kluwer Academic Publishers, Dordrecht and Boston and London, 1991. ISBN: 0792314727.

R. Salakhutdinov and G.E. Hinton. Deep Boltzmann Machines. In *Proceedings of the Artificial Intelligence and Statistics*, pages 448–455, 2009.

M. San Biagio, M. Crocco, M. Cristani, S. Martelli, and V. Murino. Heterogeneous Auto-similarities of Characteristics (HASC): Exploiting Relational Information for

Classification. In *Proceedings of the IEEE International Conference on Computer Vision*, pages 809–816, 2013.

I.M. Sintorn, M. Homman-Loudiyi, C. Söderberg-Nauclér, and G. Borgefors. A Refined Circular Template Matching Method for Classification of Human Cytomegalovirus Capsids in TEM Images. *Computer Methods and Programs in Biomedicine*, 76(2):95–102, 2004.

V. Vapnik. *The Nature of Statistical Learning Theory*. Springer, New York, 1995. ISBN: 0-387-94559-8.

R. Wang, H. Guo, L.S. Davis, and Q. Dai. Covariance Discriminative Learning: A Natural and Efficient Approach to Image Set Classification. In *Proceedings of the IEEE Conference on Computer Vision and Pattern Recognition*, pages 2496–2503. IEEE, 2012.

Z. Wen, Z. Li, Y. Peng, and S. Ying. Virus Image Classification Using Multi-scale Completed Local Binary Pattern Features Extracted from Filtered Images by Multi-scale Principal Component Analysis. *Pattern Recognition Letters*, 79:25–30, 2016.

Z. Zhang, M. Wang, and A. Nehorai. Optimal Transport in Reproducing Kernel Hilbert Spaces: Theory and Applications. *IEEE Transactions on Pattern Analysis and Machine Intelligence*, 2019.

8

Neural Network Optimizers for Brain Tumor Image Detection

T. Kalaiselvi and S.T. Padmapriya

Department of Computer Science and Applications, The Gandhigram Rural Institute - Deemed to be University, Tamilnadu, India

8.1 Introduction

Brain tumors are caused by the unnecessary development of abnormal cells within the brain. The brain tumor may be benign or malignant. Benign tumors are noncancerous, whereas malignant tumors are cancerous (Thiruvenkadam and Perumal, 2016). The significance of brain tumors in India ranges from 5 to 10 per 100 000 people (for Health Informatics, 2016). It is essential to detect brain tumors at an early stage. In the proposed work, we classified abnormal slices from MRI images of human head scans. MRIs are more advantageous than CT or x-Ray imaging technologies (Sriramakrishnan et al., 2019). There are four types of MRI images: T1-weighted, T2-Weighted, T1C, and FLAIR (Kalaiselvi et al., 2016). They produce voluminous 3D images, which help identify the exact location of the tumor. The orientations of brain MRIs are axial, coronal, and sagittal. We used T2-weighted images in the axial view to train and test the neural network model in the proposed work.

The human brain mimic is an artificial neural network (ANN). According to a specific architecture, a neuron in ANN is a mathematical function that collects and classifies information. It is a combination of the input, hidden, and output layers. CNN is a deep neural network model used for computer vision tasks (Abd-Ellah et al., 2018). CNN consists of input, convolutional, pooling, fully connected, and output layers. We have used both a simple neural network model and a convolutional neural network model for the proposed work experiment.

Optimizers play a vital role in neural networks. Optimizers can be defined as algorithms or methods that intend to modify the neural network's attributes to minimize the losses (Duchi et al., 2011). The characteristics of neural networks include weights and learning rate. There are different types of optimizers in neural networks. They are stochastic gradient descendant (SGD), adaptive moment estimation (adam),

Corresponding Author: Author; kalaiselvi.gri@gmail.com, stpadmapriya@gmail.com

Applied Smart Health Care Informatics: A Computational Intelligence Perspective, First Edition.
Edited by Sourav De, Rik Das, Siddhartha Bhattacharyya, and Ujjwal Maulik.

adaptive gradient algorithm (adagrad), root mean square propagation (rmsprop), adamax, adadelta, and Nesterov momentum (nadam). The optimizers define the modification of weights and learning rates in the neural network to reduce losses.

8.2 Related Works

Mohsen et al. (2018) developed a classification methodology for brain tumor classification from brain MRI volumes. The discrete wavelet transform (DWT) technique was used to extract features. They employed the Fuzzy c means algorithm for image segmentation. These extracted features were trained to the system and classified using deep neural networks (DNN). The authors concluded that CNN is the new methodology for classification, but it takes more time to process than large scale images.

Hwang et al. (2019) used batch normalization (BN) in a convolution block for faster convergence. The process of normalizing the activation function or the output of the convolution layer is known as BN. To avoid overfitting, dropout is used in the final layer. Dropout is a regularization technique. The authors also mentioned that their model complies with the training data but not the testing data. Kamnitsas et al. (2017) developed an 11-layered, 3D CNN for the segmentation of brain lesions. In the proposed models, we used different optimizers in simple artificial neural network models and the CNN model for brain tumor classification. Kamnitsas et al. (2017) used the BraTS and ISLES data sets for their proposed work. We also used the BraTS data set for our classification task.

Kleesiek et al. (2016) proposed a 3D CNN, a deep learning architecture for brain portion extraction from MRI images. They used T1-weighted images for their experiments and clinical data sets from brain tumor patients. In the proposed work, T2-weighted images were used for the classification of brain tumors. We also used both the BraTS and clinical data sets for our research. Pereira et al. (2016) developed a segmentation method that is based on CNN. They used 3 x 3 kernels for convolution. In the proposed work, we also used 3 x 3 kernels. If small kernels are used in convolution, it is possible to design a deeper architecture, which reduces overfitting.

Zhao and Jia (2016) proposed an automatic segmentation method to detect brain tumors. This segmentation method is based on CNNs. They described exploring several CNN architectures to utilize the self-learning property further in their future work. In our proposed work, we have proposed several CNN models and analyzed which model is best suitable for brain tumor classification. Kavitha et al. (2016) developed a method for brain tumor segmentation. They used a genetic algorithm with a support vector machine classifier. SVM classifiers are am ML classification technique. Here, every image feature should be extracted and trained to the system, a hectic process. In our proposed method, we used CNN for image classification, where the system trains itself. (Suhag and Saini, 2015) proposed brain tumor detection and classification using an SVM classifier. In our proposed method, we used CNN to classify and detect brain tumor slices from a set of MRI volumes. SVM classifier is an ML algorithm, whereas CNN is a deep learning algorithm. CNN is a self-learning algorithm in which the system trains itself using filters during convolution.

8.3 Background

8.3.1 Types of Neural Networks

There are different types of neural networks that are differentiated by their outcome and architecture. Some major types include perceptron, CNN, recurrent neural network, long short term memory, gated recurrent units, Hopfield network, Boltzmann machine, deep belief networks, autoencoders, and generative adversarial networks. Perceptron belongs to ANN's. It can handle binary inputs and outputs using sigmoid activation. In perceptron, if the weighted sum is lower or equal to the threshold value, then it produces "0" as output, whereas if the weighted sum is larger than the threshold value, then "1" is the output. CNNs are exclusively designed and developed for computer vision. These neural networks are used for comprehensive image recognition and classification. Some of the primary operations performed in CNN are convolution operation, nonlinearity, pooling or subsampling, and classification using fully connected layers. Recurrent neural networks (RNNs) are more suitable for dealing with time-series data. They are capable of recognizing the sequential characteristics of data. Feedback loops are used in RNNs to loop information back to the network. This quality allows RNNs to deal with sequential, time series, and temporal data. Some of the significant applications of RNNs are natural language processing and speech recognition.

Long short term memory (LSTM) is an enhancement of RNNs. RNNs suffers from the vanishing gradient problem. LSTM networks are designed with a memory cell that can hold the memory for long periods and are capable of dealing with long term dependencies. LSTM uses inputs and forgets gates, which allow the gradient to flow better, thus decreasing the vanishing gradient problems. Gated recurrent units (GRU) can reduce the vanishing and exploding gradient problem caused by RNN. GRU does not maintain an internal cell state. Instead of the inner cell state (used for internal storage) in LSTM, GRU incorporates the hidden state. The information stored in a hidden condition is transferred to the next gated recurrent unit. A Hopfield network is the collection of perceptron that can get rid of Exclusive OR (XOR) problems. Here, the neurons will transmit the signals back and forth in a closed feedback loop. Initially, Hopfield networks were designed to store different patterns and memories. But it can recognize any of the learned patterns by uncovering the corrupted data about that pattern. It is capable of restoring the closest pattern of the corrupted data. Networks of neurons connected symmetrically to make technical decisions in a neural network are known as the Boltzmann machine. It includes visible units (i.e. input and output) and hidden units in various layers. The Boltzmann machine enables faster learning using one layer of feature detectors at a time. The deep belief network (DBN) varies from other deep neural networks. In deep neural networks, there are connections between all neurons across layers, whereas DBNs have connections from one layer to another. The neurons will not have any connections between them. A DBN can reconstruct its inputs. The layers in a DBN act as feature detectors. DBNs are employed for classification tasks. Autoencoders are used for dimensionality reduction. They are capable of learning efficient data in an unsupervised method. After dimensionality reduction, autoencoders reconstruct from the encoding representation an output very close to its original input. Autoencoders are also used for denoising. An approach being used for generative modeling is known as generative adversarial networks (GANs). This is a supervised learning method with two sub-models named the generator and discriminator models. The generator model creates new examples similar to the training data. The discriminator model is employed to classify the samples as either real or spurious. The generator and discriminator models are trained together in an adversarial manner until the discriminator model fails to identify the fake examples. This shows that the generator model can produce standards closer to the original input.

8.3.2 Tunable Elements of Neural Networks

Tunable neural networks are characterized by the necessary parameters and hyperparameters of neural networks, regularization techniques, and optimization techniques used in neural networks. The basic parameters, hyperparameters, regularization, and optimization techniques are discussed in detail below.

8.3.2.1 Basic Parameters

Parameters can be defined as the training data properties that learn on their own during the training process. In a neural network, the basic parameters are the weights and the bias. The basic units of a neural network are the neurons, which are connected from layer to layer. The importance and bias are added to the inputs during the transmission of information between the neurons. The output of hidden layers in neural networks can be defined as a summation of weights and input, which are finally added to bias. Weights determine the amount of influence that inputs have on the outputs. Biases will be a constant added as a unit in each layer. Typically, the value of bias will be set to "1". Bias will not have any incoming connections, but it will have outgoing links. They will have their weights, and they will not have any influence from the previous layers.

8.3.2.2 Hyperparameters

Hyperparameters of neural networks can be defined as the neural network's properties that administer the entire training process. Some of the model hyperparameters are described below. Hyperparameters have direct control over the training algorithm's behavior as well as the model's performance during training. To build a successful architecture of the neural network, it is necessary to choose appropriate hyperparameters. For example, if we choose a learning rate that is too low, the model will not learn the basic patterns of the data. If the learning rate is set high, then it will have some collisions.

The learning rate must be chosen as an optimal value. If the learning rate is too low compared to the optimal value, then the training time will be longer (i.e. it needs hundreds or thousands of epochs to train the model). If the learning rate is more significant than the optimal value, then the model will not converge. The optimal value of the learning rate is 0.001.

Mini batch size is advantageous in preventing the training process from halting at its local minima. An optimal value for the mini-batch size is 32. If the chosen mini-batch size is more extensive, then it requires more memory for the training process.

The number of epochs plays a more significant role in training the neural network models. The right number of epochs can be decided based on the validation error metric. Here, early stopping criteria can be applied to determine the number of epochs needed to train the model. If early stopping criteria are defined, it will stop the training process if the validation error does not decrease for the last 10 to 20 epochs.

The number of hidden units measures the model's learning capacity. To make a neural network learn a simple function, only a few hidden units are needed. If the complexity of the function increases, the model will require more learning capacity.

8.3.2.3 Regularization Techniques

Regularization techniques are used to generalize the models by applying minimal modifications to the learning algorithms. By using regularization techniques, the performances of the models are increased. The types of regularization techniques used in deep neural networks are L2 and L1 regularization, dropout, data augmentation, and early stopping.

The most common types of regularization are L1 and L2 regularization. These techniques tend to make an update process, and there will be a decrease in the values of weight matrices, which reduces the overfitting of the model. Dropout is also one of the most commonly used regularization techniques. Dropouts can be applied both in the input and hidden layers. This process involves the selection of some random neurons and removes them. It means that all incoming and outgoing connections of the neuron are also removed. Dropout is used as a hyperparameter in neural networks. The probability of deciding the number of nodes dropped from the network can be defined as the dropout function's hyperparameter. It also avoids overfitting of the model.

Data augmentation is another popular hyperparameters of neural networks. Data augmentation is defined as the shearing, rotation, and flipping of images. Data augmentation is a process that depends on the data. Data augmentation can be used as a hyperparameter in the case of inadequate training data samples.

One of the most popular regularization techniques to avoid overfitting is the early stopping criteria. This method performs the process of learner updating to fit the training data in each iteration. An early stopping criterion is defined based on the value of loss or the value of accuracy. This method stops the training process if the accuracy does not improve over the prescribed iterations. It can also be stated for early stopping if the value of loss increases.

8.3.2.4 Neural Network Optimizers

Optimization algorithms are used for maximizing or minimizing an objective function. In the case of neural networks, optimization algorithms are used to reduce the error function (Kingma and Ba, 2014; Tieleman and Hinton, 2012; Wilson et al., 2018). To train a neural network model efficiently, the internal parameters of a model are used. These internal parameters are modified using optimization strategies and algorithms to produce optimum values and accurate results. The optimization algorithm can be classified into two major categories: first-order and second-order optimization algorithms.

First-order optimization algorithms are intended to maximize or minimize a loss function. This is accomplished by using the value of its gradients for the parameters. An example of a first-order optimization algorithm is the gradient descent algorithm. The first-order derivative shows the increasing or decreasing function at a particular point. Second-order methods (Hessian) are also used to maximize or minimize the loss function. The matrix of second-order partial derivatives is known as Hessian. The second-order partial derivative provides a quadratic surface that touches the curvature of the error surface.

SGD is one of the fastest techniques that performs a parameter update for each training example. Due to frequent updates, the parameter updates will have a high variance. It leads to fluctuations in the loss function to different intensities. The disadvantage of SGD is that it produces complications in the convergence to the exact minimum due to frequent updates and changes. The variance in parameter updates is reduced by a Mini batch gradient descent, which lead to stable convergence.

Adagrad is well suited for sparse data since it allows the learning rate to adapt according to the parameters. It creates significant updates for infrequent parameters and smaller updates for frequent parameters. The advantage of adagrad is not needing to tune the learning rate manually. The main disadvantage of adagrad is that the learning rate tends to decrease or decay.

Adadelta is an extension of adagrad, which overcomes the problem of learning rate decay. In adadelta, there is no need to set a default learning rate. Adam manipulates adaptive learning rates for each parameter; it consists of an exponentially decaying average of past gradients that are similar to momentum. When compared to other optimizers, adam performs well due to its faster convergence, efficiency, etc. It also rectifies the vanishing learning rate problem, high variance in parameter updates, and fluctuating loss functions.

Rmsprop optimizer is very similar to SGD with momentum. It restricts the oscillations in the vertical direction, which increases the learning rate and larger steps for faster convergence. Adamax uses the normalization values. If infinity norm is used, adamax performs well and becomes stable. Nesterov momentum (Nadam) makes a big jump based on the previous momentum, and then the gradient is calculated. After that, a parameter update is done.

8.4 Case Study - Brain Tumor Detection

8.4.1 Methodology

A simple neural network and CNN model were used in our proposed work to conducting our experiments. The architecture of the neural network and CNN models are shown in Figures 8.1 and 8.2, respectively. The simple neural network model consists of an input layer, a hidden layer, and an output layer. The CNN model is designed with five layers, and each contains convolution and a pooling layer. We have used batch normalization to improve the speed, performance, and stability of the neural networks. Batch normalization adjusts and scales the activation functions for normalizing the input layer. Here the default value of epochs is set to 30. We used an early stopping criterion to minimize the training time of the neural network model. In the proposed work, various optimizers were used to check the accuracy of neural network models. We conducted an analytical study on the optimizers and chosen the best among them.

8.4.2 Data Sets and Metrics

The BraTS 2013 data set (BraTS2013 Challenge, MRI brain tumor database) (BR) and a sample clinical data set from the Whole Brain Atlas (WBA) (Johnson and Becker, 1999) were used to train and test the proposed models. Tensorflow is an open-source software library developed by the Google Brain Team. The second-generation system of Google is used to implement and deploy several machine learning models. Dataflow graphs are used to represent the computations of Tensorflow. Keras is an open-source library written in Python. It runs as front-end software for Tensorflow and Theano. Francois Chollet, a Google engineer, developed keras. It provides a higher-level programming interface wrapper for the lower-level application programming interface. It handles designing models, defining the layers, and setting up multiple input and output models. Some of the popular evaluation metrics used for image classification tasks are accuracy, false alarm, and missed alarm. We used the below equations to evaluate our experiments.

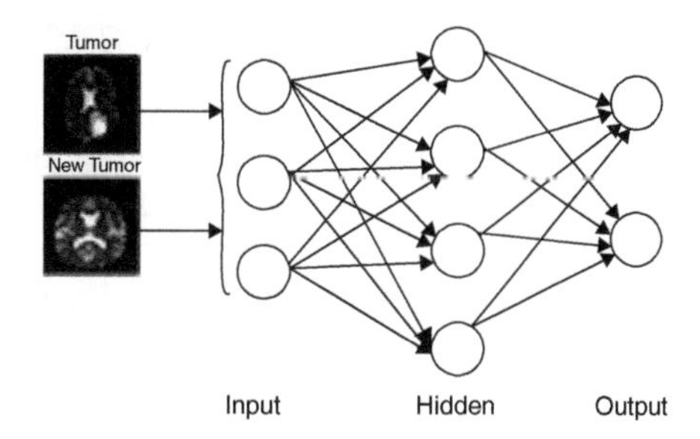

Figure 8.1 Architecture of a simple neural network.

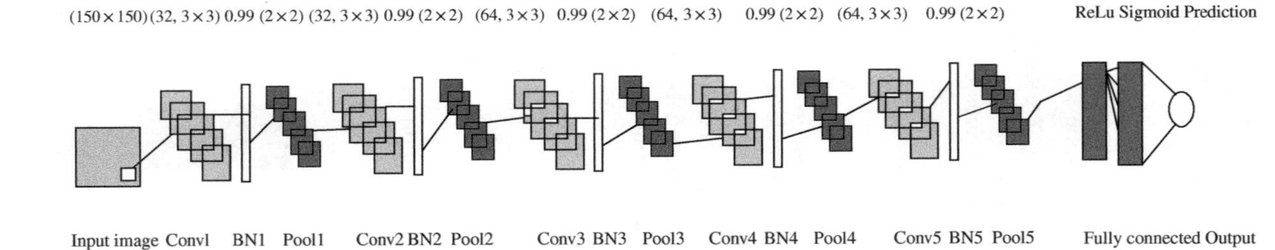

Figure 8.2 Architecture of a convoluted neural network model.

$$\% \textit{ Missed Alarm } (\text{MA}) = \left(\frac{\text{FP}}{\textit{Total number of slices}}\right) \times 100 \tag{8.1}$$

$$\% \textit{ False Alarm } (\text{FA}) = \left(\frac{\text{FN}}{\textit{Total number of slices}}\right) \times 100 \tag{8.2}$$

$$\textit{Accuracy} = \frac{(\text{TP} + \text{TN})}{(\text{TP} + \text{TN} + \text{FP} + \text{FN})} \tag{8.3}$$

8.4.3 Results and Discussion

We conducted our experiments on both simple neural network and convolutional neural network models. The architectures of these models are depicted in Figures 8.1 and 8.2. Initially, we experimented on a simple neural network model by varying the optimizers. In the beginning, the model was trained for 10 epochs. When we used a stochastic gradient descent optimizer, the accuracy of the model was 79%. Similarly, the rmsprop optimizer accuracy was 84.38%. For the adagrad optimizer, the accuracy was 77.82%. For adadelta, the accuracy was 86.08%. For adamax, the accuracy was 96.76%. For nadam, the accuracy was 87.97%. Finally, the adam optimizer got an accuracy of 97.4%.

Later, the simple neural network was trained for 20 epochs, and the accuracy was recorded as the optimizers were varied. The SGD optimizer produced an accuracy of 83.52%. Rmsprop had an accuracy of 84.38%. Adagrad produced an accuracy of 77.82%. Adadelta produced an accuracy of 86.08%. Adamax produced an accuracy of 96.76%. Nadam had an accuracy of 87.97%. Finally, as specified earlier, the adam optimizer outperformed all other optimizers, and it produced a 98.94% accuracy. In simple neural network models, the adam optimizer outperforms all other optimizers because of its sparsity. Adam also combines the advantages of rmsprop and adagrad, which are the two extensions of stochastic gradient descent optimizers. The performances of all the optimizers in the simple neural network model are shown in Figure 8.3. The accuracy of all optimizers is listed in Table 8.1.

The architecture of the CNN model is shown in Figure 8.2. In this model, we used the early stopping criterion, which reduced the model's training time. The total number of assigned epochs is 30. The performance of all optimizers using the early stopping criterion is shown in Table 8.2.

When we used the SGD optimizer, it produced an 82% accuracy, and the model stopped training at the fourteenth epoch. While using the rmsprop optimizer produced an 89.25% accuracy, and the training process stopped at the sixteenth epoch. The adagrad optimizer produced an accuracy of 79.61%, and the model stopped training at the twentieth epoch. When we used the adadelta optimizer, it produced an 88.12% accuracy, and the model stopped training during the thirteenth epoch. While using the adamax optimizer, the accuracy obtained was 97.8%, and the model stopped training after the thirteenth epoch. The nadam optimizer gave an accuracy of 89.10%, and the training process stopped after the twenty-first epoch. Finally, we used the adam optimizer and obtained a 98.7% accuracy; the training process stopped after the thirteenth epoch. The performance of optimizers in the CNN model is depicted in Figure 8.4. The number of training

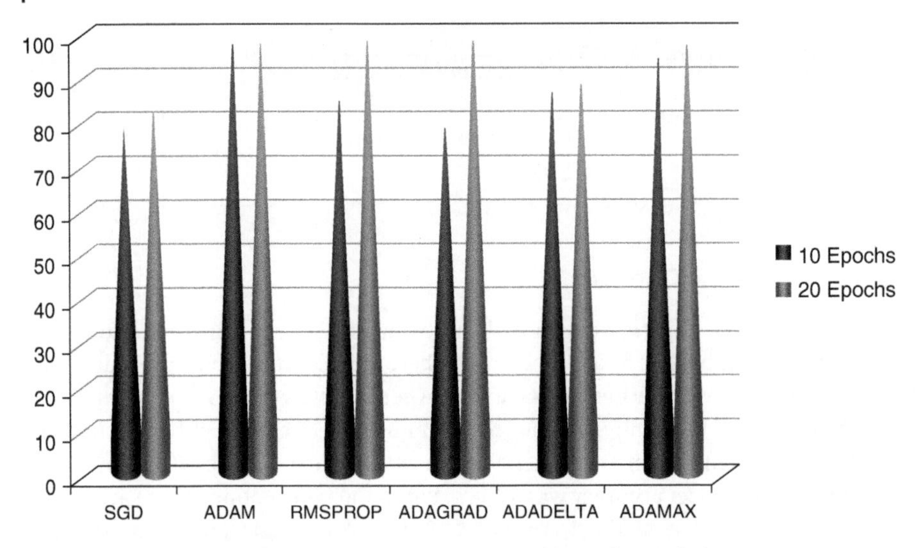

Figure 8.3 Performance of implemented networks with varying optimizers.

Table 8.1 Accuracy of artificial neural network models for differing optimizers.

S Number	Optimizer	Accuracy (%, 10 Epochs)	Accuracy (%, 20 Epochs)
1	SGD	79	83.52
2	Adam	97.4	98.94
3	Rmsprop	84.38	97.5
4	Adagrad	77.82	97.8
5	Adadelta	86.08	87.6
6	Adamax	96.76	97.71
7	Nadam	87.97	97.4

Table 8.2 Performance analysis of neural network optimizers using early stopping criterion.

S Number	Optimizer	Accuracy (%)	Number of Epochs
1	SGD	82	14
2	Adam	98.7	13
3	Rmsprop	89.25	16
4	Adagrad	79.61	20
5	Adadelta	88.12	13
6	Adamax	97.8	13
7	Nadam	89.10	21

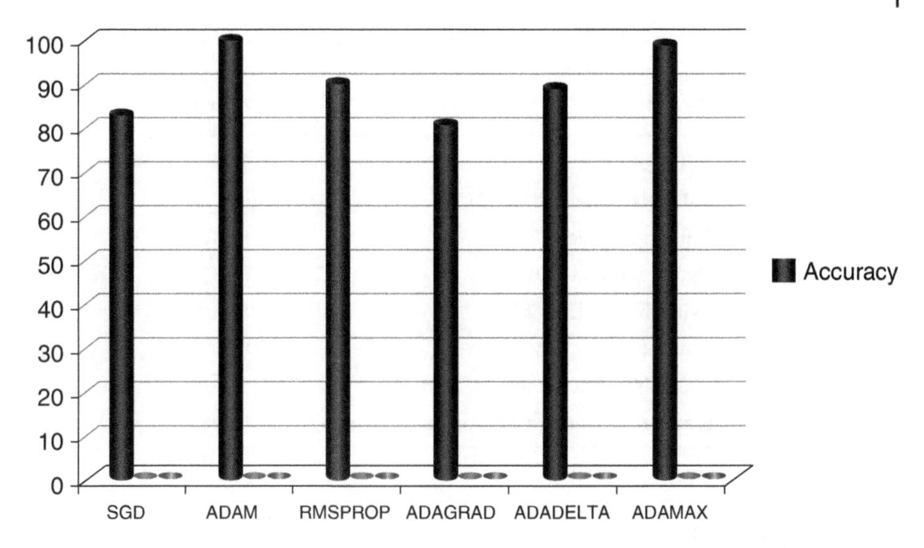

Figure 8.4 Accuracy obtained by optimizers in a convolutional neural network.

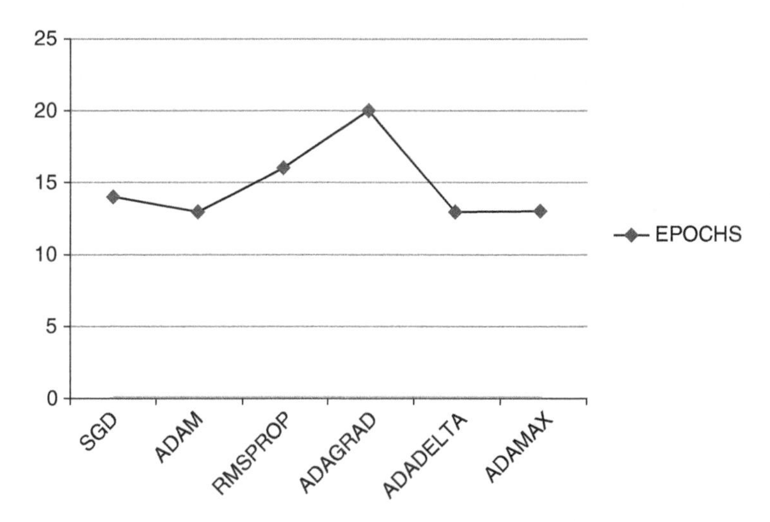

Figure 8.5 Number of epochs required for training by CNN optimizers with early stopping criterion.

epochs taken by all optimizers is shown in Figure 8.5. Compared with the simple neural network model, the training time was gradually reduced using the early stopping criterion in CNN. The performance of the proposed method was compared with other state-of-the-art methods (Kalaiselvi et al., 2020) and are described in Table 8.3.

Table 8.3 Comparison of the proposed method with state-of-the-art methods.

S number	Method	Accuracy (%)
1	PCB (2011)	58.9
2	Kalaiselvi's method (2012)	90
3	Statisical method (2014)	85.8
4	Hybrid method (2015)	88.3
5	Avinash's method (2015)	87.3
6	Wavelet method (2015)	89.6
7	FLSCBN method (2020)	88.91
8	Proposed method (using Adam optimizer)	**98.94**

8.5 Conclusion

In the proposed work, we conducted an analytical study of optimizers in a simple neural network and CNN models. From these results, we concluded that the adam optimizer performed well compared to SGD, rmsprop, adagrad, adamax, adadelta, and nadam optimizers. The adam optimizer model achieved an accuracy of 98% in tumor image detection from the MRI of human brain scans. In the future, we pursue developing a novel optimizer for the neural network model to be tested on abnormal slice classification of human MRI head scans.

References

BRATS 2015 challenge, MRI brain tumor database. URL https://www.smir.ch/BRATS/Start2015. (accessed on 23.8.2020).

M.K. Abd-Ellah, A.I. Awad, A.A.M. Khalaf, and H.F.A. Hamed. Two-phase multi-model automatic brain tumour diagnosis system from magnetic resonance images using convolutional neural networks. *EURASIP Journal on Image and Video Processing volume*, 2018(1), September 2018. doi: 10.1186/s13640-018-0332-4.

J. Duchi, E. Hazan, and Y. Singer. Adaptive subgradient methods for online learning and stochastic optimization. *Journal of Machine Learning Research*, 12(null):2121–2159, July 2011. ISSN 1532-4435.

Centre for Health Informatics, 2016. URL www.nhp.gov.in. (accessed on 02 May 2020).

H. Hwang, H.Z.U. Rehman, and S. Lee. 3D U-Net for skull stripping in brain MRI. *Applied Sciences*, 9(3):569, February 2019. doi: 10.3390/app9030569.

K.A. Johnson and J.A. Becker. The whole brain atlas (WBA), 1999. URL https://www.med.harvard.edu/aanlib/. (accessed on 20.8.2020).

T. Kalaiselvi, P. Sriramakrishnan, and K. Somasundaram. Brain abnormality detection from MRI of human head scans using the bilateral symmetry property and histogram similarity measures. IEEE, December 2016. doi: 10.1109/icsec.2016.7859867.

T. Kalaiselvi, S.T. Padmapriya, P. Sriramakrishnan, and K. Somasundaram. Deriving tumor detection models using convolutional neural networks from MRI of human brain scans. *International Journal of Information Technology*, 12(2):403–408, February 2020. doi: 10.1007/s41870-020-00438-4.

K. Kamnitsas, C. Ledig, V.F.J. Newcombe, J.P. Simpson, A.D. Kane, and et al. Efficient multi-scale 3d CNN with fully connected CRF for accurate brain lesion segmentation. *Medical Image Analysis*, 36:61–78, February 2017. doi: 10.1016/j.media.2016.10.004.

A.R. Kavitha, L. Chitra, and R. Kanaga. Brain tumor segmentation using genetic algorithm with SVM classifier. *International Journal of Advanced Research in Electrical and Electronics Engineering*, 5(3):1468–1471, March 2016. doi: 10.15662/IJAREEIE.2016.0503043.

D.P. Kingma and J. Ba. Adam: A method for stochastic optimization. 2014.

J. Kleesiek, G. Urban, A. Hubert, D. Schwarz, K. Maier-Hein, and et al. Deep MRI brain extraction: A 3D convolutional neural network for skull stripping. *Neuroimage*, 129:460–469, April 2016. doi: 10.1016/j.neuroimage.2016.01.024.

H. Mohsen, E.S.A. El-Dahshan, E.S.M. El-Horbaty, and A.B.M. Salem. Classification using deep learning neural networks for brain tumors. *Future Computing and Informatics Journal*, 3(1):68–71, June 2018. doi: 10.1016/j.fcij.2017.12.001.

S. Pereira, A. Pinto, V. Alves, and C.A. Silva. Brain tumor segmentation using convolutional neural networks in mri images. *IEEE Transactions on Medical Imaging*, 35(5):1240–1251, 2016. doi: 10.1109/TMI.2016.2538465.

P. Sriramakrishnan, T.m Kalaiselvi, K. Somasundaram, and R. Rajeswaran. A rapid knowledge-based partial supervision fuzzy c-means for brain tissue segmentation with CUDA-enabled GPU machine. *International Journal of Imaging Systems and Technology*, 29(4):547–560, May 2019. doi: 10.1002/ima.22335.

S. Suhag and L.M. Saini. Automatic brain tumor detection and classification using SVM classifier. *International Journal of Advances in Science, Engineering and Technology (IJASEAT)*, 3(4):119–123, 2015.

K. Thiruvenkadam and N. Perumal. Fully automatic method for segmentation of brain tumor from multimodal magnetic resonance images using wavelet transformation and clustering technique. *International Journal of Imaging Systems and Technology*, 26(4):305–314, December 2016. doi: 10.1002/ima.22202.

T. Tieleman and G. Hinton. Lecture 6.5-rmsprop: Divide the gradient by a running average of its recent magnitude., 2012.

A.C. Wilson, R. Roelofs, M. Stern, N. Srebro, and B. Recht. The marginal value of adaptive gradient methods in machine learning. 2018.

L. Zhao and K. Jia. Multiscale CNNs for brain tumor segmentation and diagnosis. *Computational and Mathematical Methods in Medicine*, 2016:1–7, 2016. doi: 10.1155/2016/8356294.

9

Abnormal Slice Classification from MRI Volumes using the Bilateral Symmetry of Human Head Scans

N. Kalaichelvi, T. Kalaiselvi*, and K. Somasundaram**

Department of Computer Science and Applications, The Gandhigram Rural Institute- Deemed to be University, Tamilnadu, India

9.1 Introduction

9.1.1 MRIs of the Human Brain

Among all vertebrates, the human brain is the largest relative to the size of the body. It weighs about 2% of a human's body and is kept highly protected by the scalp, skull and three layered meninges. It is surrounded by a liquid called cerebrospinal fluid (CSF), which acts as a cushion to provide basic protection to brain inside the skull. The three major regions are the cerebrum, cerebellum, and the brainstem. The cerebrum occupies 85% of the brain portion. It contains about 86 billion nerve cells called neurons in the gray matter (GM) and billions of nerve fibres called axons and dentrites in the white matter (WM). These neurons are connected by synapses with trillions of connections. The cerebrum is divided into two halves called the left hemisphere (LHS) and right hemisphere (RHS), which are separated by a longitudinal fissure called the inter hemispheric fissure (IHF). The hemispheres are symmetric in nature, based on the IHF in a normal brain as shown in Figure 9.1. These hemispheres have control over the opposite side of the body; the LHS is responsible for the functioning of the right side of the body and vice versa.

Of the available brain imaging techniques, magnetic resonance imaging (MRI) is considered non-invasive since it uses radio waves and a high magnetic field instead of radiation to produce more informative 3-dimensional (3D) images of the cerebrum, cerebellum and brain stem. The device contains a large doughnut-shaped tunnel with a built-in magnet. A rotating table slides into the magnetic tunnel to carry the patient to be scanned. During imaging, radio waves are used to accelerate the body atoms. The energy released by these atoms are captured by a sensor and sent to the computer-aided imaging system that produces 3D images of the scanned area. The MRI produces three orientations: axial, coronal, and sagittal. The axial images are taken with a bird's eye-view from top to bottom. The sagittal images are taken

*kalaiselvi.gri@gmail.com, chelvi.kalai7@gmail.com, ka.somasundaram@gmail.com

Applied Smart Health Care Informatics: A Computational Intelligence Perspective, First Edition.
Edited by Sourav De, Rik Das, Siddhartha Bhattacharyya, and Ujjwal Maulik.

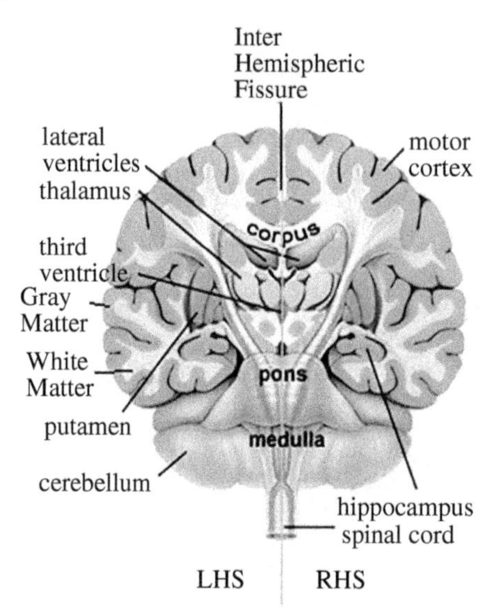

Figure 9.1 Anatomy of the human brain.

from left ear to right ear, and the coronal images are taken from front to back. Thus, the MRI can be used to detect or diagnose brain abnormalities like cysts, tumors, hemorrhage, swelling, inflammation or clots in blood vessels, developmental and structural abnormalities, infections and damage to the brain by an injury or stroke, trauma, etc. MRIs can provide detailed pictures of parts of the brain to diagnose problems in areas that can't be seen well in other scans like CTs or ultrasounds. These are the T1-weighted (T1), T1-contrast enhanced (T1c), T2-weighted (T2) and fluid-attenuated inversion recovery (FLAIR) images. The combination of all these sequences is called a multimodal MRI (MMMRI). To analyse healthy tissues, T1 images are used. It contains a high intensity region for WM, low intensity regions for GM, and a dark region for CSF. To analyse pathological tissues, T2 images are used. The intensity arrangements of T2 images are that CSF remains a high intensity region, GM as a low intensity, and WM as a dark region. T1c images are taken by injecting a contrasting agent into the patient, which creates a fluorescent signal on the tumor region, if present. In a FLAIR sequence (similar to the T2 sequence), the signals from the fluid are suppressed into a void, thus tumors have brighter regions for tumors than other tissue regions. It is mainly used to extract the tumor portion (Thiruvenkadam, 2011; Kalaiselvi and Selvi, 2017). Sample multimodal MRI images with tumor regions are shown in Figure 9.2.

9.1.2 Normal and Abnormal Slices

A normal brain is symmetric in nature about the vertical line passing through, the inter hemispheric fissure (IHF). The IHF bisects the human body passes along the mid-sagittal plane (MSP). Thus, the bisected pieces of brain, based on the MSP, are called cerebral hemispheres. Brain abnormalities are any atypical feature in its structure or functioning. Certain brain abnormalities such as dementia, stroke, hematomas, blood clots, cerebral edema, brain tumor, etc. cause damage, swelling,

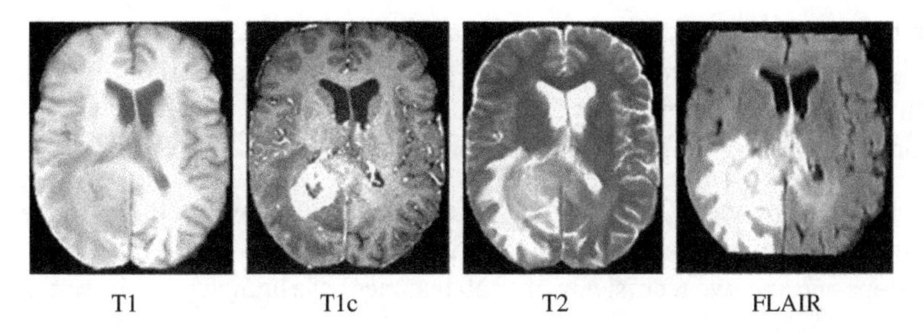

Figure 9.2 MRI of a brain tumor.

or displacement of the brain. This affects the bilateral symmetric nature of the brain. Brain tumors majorly disturb the bilateral symmetry since they grow abnormally inside brain or arose in another organ and metastasized to the brain. Since they are a higher intensity, they are explicitly visible in MRI images, especially in T2, T1c, and FLAIR sequences (Stegmann et al., 2005).

Abnormal slices are identified by experienced clinicians. However, to identify the abnormal slices from hundreds of slices is a time-consuming process. Computers aid in abnormality identification that further leads to treatment planning. In recent years, researchers have developed many algorithms for detecting brain abnormalities. In particular, machine learning models play a big role in abnormal slice classification (Karuppanagounder and Thiruvenkadam, 2009).

9.1.3 Background

Classification is the process of categorizing a pixel or an object in an image into one or more classes. It analyses various numerical and categorical features of the image and organizes them into classes based on those values. Classification models are available to work in a machine learning environment. Predictive models approximate a mapping function (f) from an input feature vector (**x**) to a discrete output class vector (**y**). Numerous classification models are used in machine learning practice such as decision trees, k-nearest neighbour (KNN), support vector machine (SVM), probabilistic-naïve Bayes, artificial neural network (ANN), back-propagation neural network (BPN), and random forest (RF).

9.1.3.1 Decision Tree Classifiers

This is a supervised learning model that is non-parametric; it aims to predict the query element by learning simple decision-based rules from the feature data. It can capture descriptive decision-making knowledge from the supplied data and learns from the data with a set of rules like if-then-else at each node. The deeper the tree, the more complex the decision rules are. There are two types of nodes: decision and leaf. Decision nodes are where decisions are made to with multiple branches. For example, a simple yes or no question may be asked at each decision node based on how the branches are split. Leaf nodes are the output nodes without any branches. Since the nodes form a tree-like structure in making decisions, these classifiers were called decision trees. Based on their application, decision trees are classification or regression trees. If the tree produces an output from a set of discrete, categorical, or predefined classes, it is a classification tree. If it produces an output as a continuous value, then it is a regression tree (Yang et al., 2003). It works based on the divide and conquer algorithm, which works as described below.

1. Select a test for root node. Create branch for each possible answer of the test.
2. Split the tree into subsets based on the instances of the test.
3. Repeat steps one and two recursively until all instances of a branch have the same class.

9.1.3.2 K-Nearest Neighbours (KNN) Classifiers

KNN is a supervised learning algorithm that performs based on the assumption that similar things exist in close proximity. It predicts the test data according to the K training data that are nearer to the test data with a greater class probability. There are six steps involved in KNN classification algorithms:

1. Initialize K as the count of nearest neighbours.
2. Calculate query data and the training samples distance.
3. Sort the training samples according to the calculated distance.
4. Determine K nearest neighbours from the sorted list.
5. Label the neighbours with their class values $\mathbf{y}$.
6. Assign the majority category $\mathbf{y}$ of neighbours to the query data.

9.1.3.3 Support Vector Machine (SVM)

An SVM is a supervised classification and prediction algorithm in machine learning. It is mostly used for classification problems. Each and every data item is plotted in an n-dimensional space where n is the number of independent variables or features ($\mathbf{x}$). The classification is performed by fixing a hyper-plane that differentiates the classes. It aims to maximize the distance between data points of distinct classes to allow confident classification of future data (Kim et al., 2012).

The dimension of hyperplanes may vary based on the number of features. The data points closer to the hyperplanes are called support vectors, shown in Figure 9.3. These support vectors fix the orientation and position of the hyperplane. To optimize

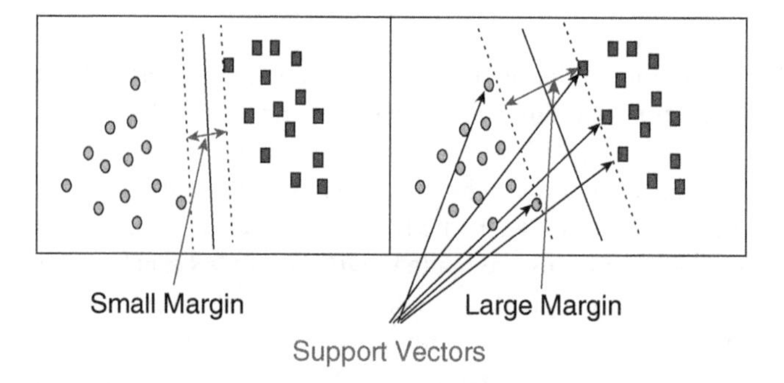

Figure 9.3 Support vector machine.

the SVM, a loss function (hinge loss) is used to maximize the margin:

$$c(\mathbf{x}, \mathbf{y}, f(\mathbf{x})) = \begin{Bmatrix} 0, \\ if\ \mathbf{y} * f(\mathbf{x}) \geq 1 \\ 1 - \mathbf{y} * f(\mathbf{x}), \\ else \end{Bmatrix} \tag{9.1}$$

When the predicted and actual values have the same sign, the loss is 0. In other cases, the loss value is calculated as described in Equation 9.1.

9.1.3.4 Naive Bayes

Naive Bayes' is a binary and multi-class classification algorithm. Based on the Bayes theorem and a prior knowledge obtained from the given data, it provides a way to calculate the probability of a hypothesis. The Bayes' theorem is

$$\Pr(A|B) = \frac{\Pr(B|A) \times \Pr(A)}{\Pr(B)}. \tag{9.2}$$

It finds the probability (Pr) of A happening, given that B has occurred. The assumption is that predictors and features are independent of each other (Zhou et al., 2015).

9.1.3.5 Artificial Neural Network (ANN)

An ANN tries to imitate the human brain by following its internal scheme of operations in decision-making processes. It is a network of connected inputs ($\boldsymbol{x}_i$) and outputs (θ_j) with weight ($\boldsymbol{w}_{ij}$) associated to each connection. It contains an input and output layer along with multiple intermediate layers as shown in Figure 9.4. Learning is done by adjusting the weights of the connections iteratively, which improves the classifier's performance. The input layer carries raw information into the network. The data are not changed in this layer. Inside the network, raw inputs are transformed by the hidden intermediate layers. The values entering the hidden layers are multiplied by their predetermined weights. The sum of these weighted inputs yields a number sent to the next layer. The output layer corresponds to the prediction of the class variable. The active nodes in the output layer produce output values.

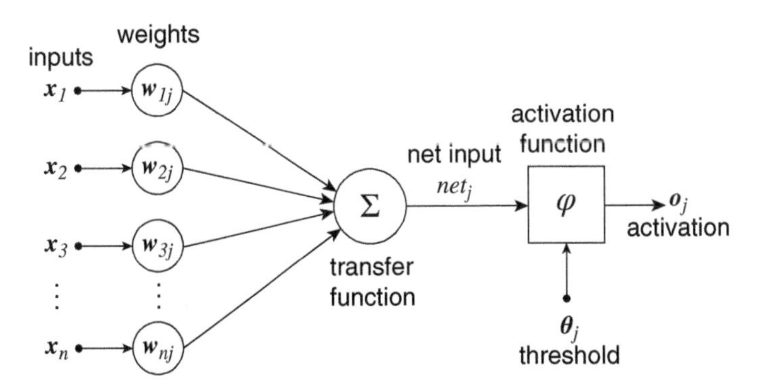

Figure 9.4 Artificial neurons.

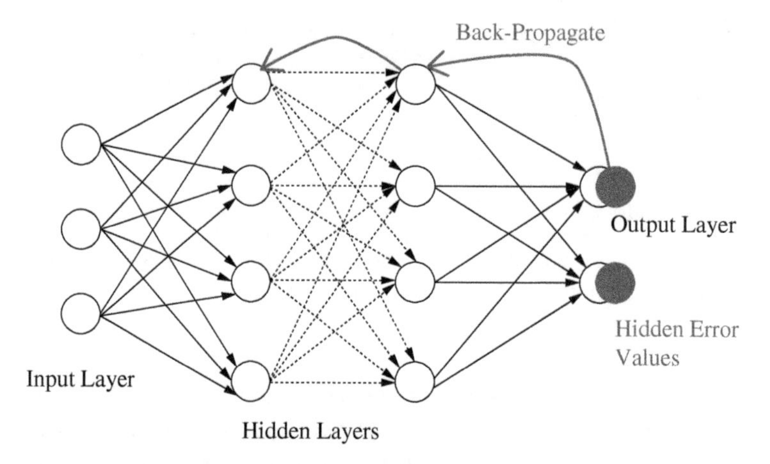

Figure 9.5 Back-propagation neural network.

The performance of the ANN model depends on the proper selection of weights (Le et al., 2012).

9.1.3.6 Back-Propagation Neural Network (BPN)

Back-propagation is a basic neural network training model. It fine tunes the weights of the inputs based on the rate of error in the previous epoch or iteration. This weight tuning reduces error and increases the reliability of the neural network model. This method helps to assess the gradient of loss function based on the weights of neurons in the network. The BPN network is shown in Figure 9.5.

1. Input neurons (X) enter through the network path.
2. Randomly selected real weights (W) are assigned to the input.
3. The output for every neuron from input layer to the hidden and output layers are calculated.
4. Error is calculated as

$$Error = ActualOutput - DesiredOutput \tag{9.3}$$

5. Traverse backwards from output layer to the hidden layer to adjust the weights (W), so that the error is reduced.
6. Repeat the above steps iteratively until the desired output is obtained.

9.1.3.7 Random Forest Classifiers

The random Forest classifier is a popular and simple machine learning classifier that uses the concept of ensemble learning and bagging to improve the prediction performance. Ensemble learning is the process of combining multiple classifiers for learning. Bagging is a technique that combines multiple classifier models to improve the quality of prediction. A schematic of a random forest classifier is shown in Figure 9.6.

1. The input data set is split into n number of training samples.
2. Decision trees are built for each training sample. These decision trees result in predictions for the test data.

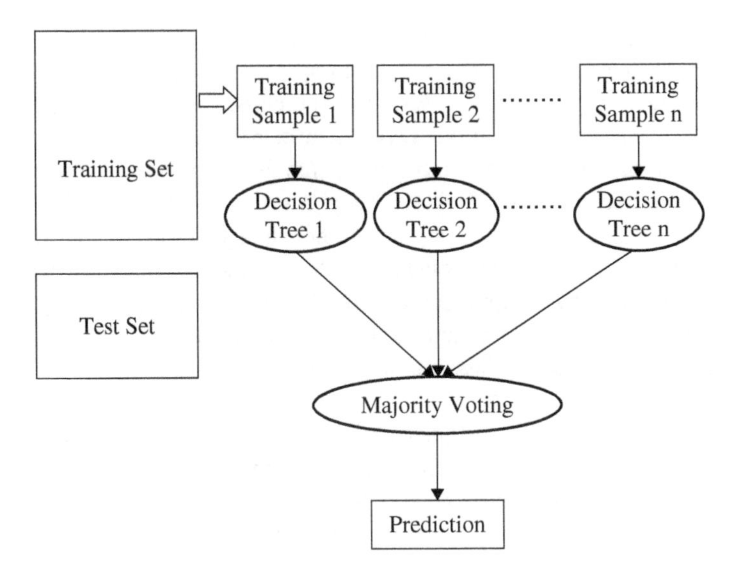

Figure 9.6 Random forest classifier.

3. Perform averaging or majority voting on the decision tree predictions for the test data set.
4. Select the most voted prediction as the final prediction for the test data.

9.2 Literature Review

Thiruvenkadam et al. (2016) proposed an ANN-based classification method to predict the abnormal slices in a volume. They bisected the image into the LCH and RCH. The histograms were taken for both images and compared with each other using 15 histogram similarity measures. A threshold value was calculated for each similarity measure with the fuzzy c means (FCM) algorithm. Based on these threshold values, the ANN was trained and tested. This model gave an accuracy of 94% for WBA, 86.4% for BraTS13, and 100% for IBSR data sets. Ramteke and Khachane (2012) proposed a four phase method starting with preprocessing the input CT image using a median filter, followed by a statistical feature extraction phase, a KNN classifier, and the extraction of the abnormal region at the postprocessing step. The accuracy of the model was compared with a kernel-based SVM classifier and produced better results.

Nader et al. (2015) proposed a technique that included sigma filtering, adaptive threshold, and tumor region detection. They combined decision tree and multi-layer perceptron classification algorithms for the learning process. T. et al. (2015) proposed a novel statistical feature to identify abnormal slices in a T2 volume. Instead of computing the mean intensity of the entire image, they removed the mean of the higher intensity pixels from the image since abnormal tissues have higher intensities. This statistical feature was used for training the model for further prediction of the abnormal slices present in a volume.

Goswami and Bhaiya (2013) proposed a method called the hybrid neuro-fuzzy system, a combination of ANN and fuzzy logic techniques. The system followed data collection, feature extraction from GLCM, and the classification of normal and abnormal images through the neuro-fuzzy system of brain images. Renjith et al. (2015) proposed an improved neuro-fuzzy technique that followed image preprocessing, wavelet decomposition, texture feature extraction from GLCM, neuro-fuzzy-based classification of normal and tumor images, and the Otsu thresholding-based tumor location identification. Megha and Sushma (2019) proposed a machine learning-based approach that skull stripped in preprocessing, segmented the tissues with thresholding and morphological operations, extracted features with GLCM, and used SVM classification to identify normal or abnormal tissues.

Jayachandran and Dhanasekaran (2013) proposed a four-stage hybrid algorithm that used statistical features and a fuzzy support vector machine (FSVM) classifier. The stages are noise reduction, feature extraction, feature reduction and classification. They achieved an accuracy of 95.80% for the normal and abnormal MRI data collected from south Indian area severity analysis and Government Medical College Hospital. Deepa and Devi (2012) proposed an abnormal slice classification algorithm using optimal texture features extracted from normal and abnormal images as well as a BPN and radial function neural network for abnormal slice classification and tumor portion segmentation to reach an accuracy of 85.7%. Soltaninejad et al. (2016) proposed a fully automatic method for abnormal slice identification in FLAIR MRI images. Novel features were calculated from intensity, Gabor textons, fractals, and curvatures of brain portions. The extremely randomized trees (ERT) classifier was used for abnormal slice identification in comparison with an SVM classifier and produced an 88% accuracy in the BRATS data set.

9.3 Methodology

In the proposed method, the input image was preprocessed by applying the bounding box technique, then split into two cerebral hemispheres (LHS and RHS) based on the MSP. Then features were extracted from each hemisphere using the histogram-based first-order, a gray level co-occurrence matrix (GLCM)-based second-order, and a gray level run length matrix (GLRLM)-based third-order (ChandraPrabha, 2019; Albregtsen, 2008). The absolute difference between the hemispheric features was taken for consideration because the greater the difference, the more likely that the image is abnormal. The feature set with 763×39 dimensions was taken into the machine learning model. The relevant features for prediction were selected using Pearson's correlation coefficient (features with correlation > 0.3). The selected features with 763×16 dimensions were obtained from the feature selection for further training and testing phases of the classification model. The KNN classifier was used in the current work to classify the MRI images into normal or abnormal. Figure 9.7 depicts the flowchart of the current method.

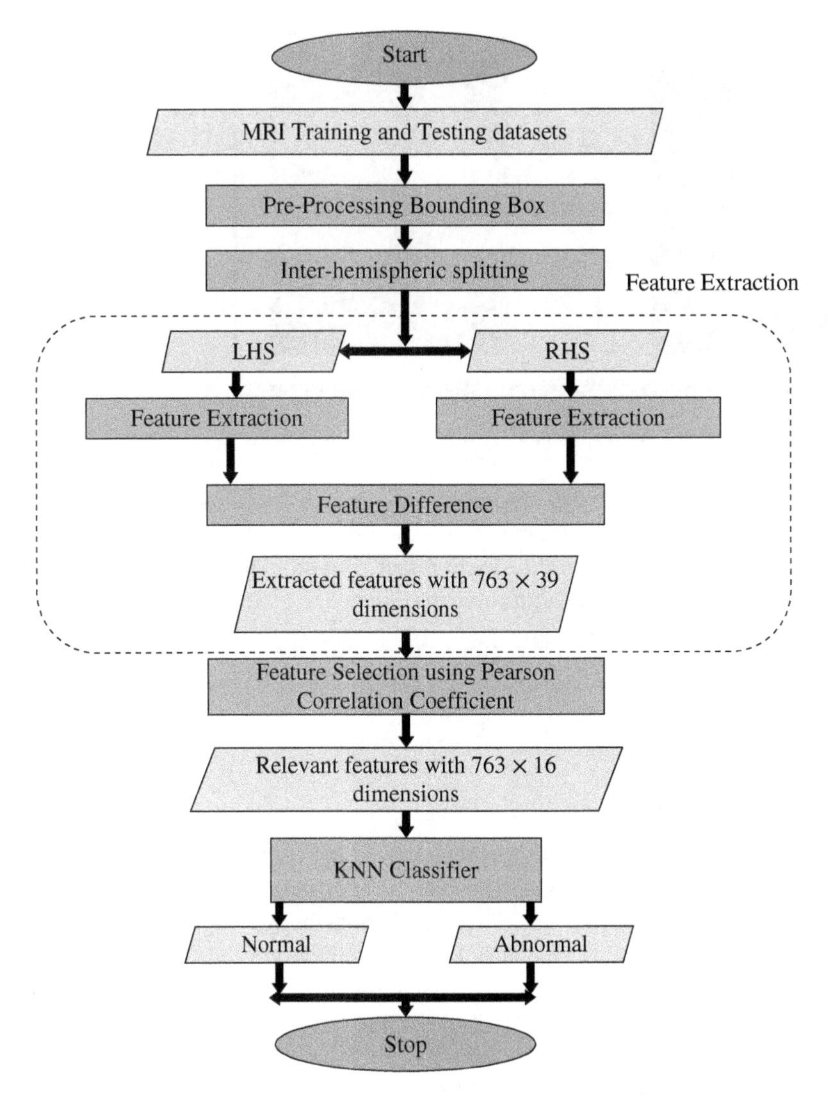

Figure 9.7 Flow chart of the proposed method.

9.3.1 Preprocessing

The background of MRI images is treated as zero. Thus, the unwanted background is cropped using a bounding box technique, which eases further processing. Here, it extracts only the brain portion as the area of interest (AOI). There are four steps involved in the bounding box technique (Thiruvenkadam et al., 2017).

1. Traverse row-wise from top to bottom in the image until non-zero data are reached. Assign the row value in variable *rowmin*.
2. Traverse row-wise from bottom to top in the image until non-zero data are reached. Assign the row value in variable *rowmax*.

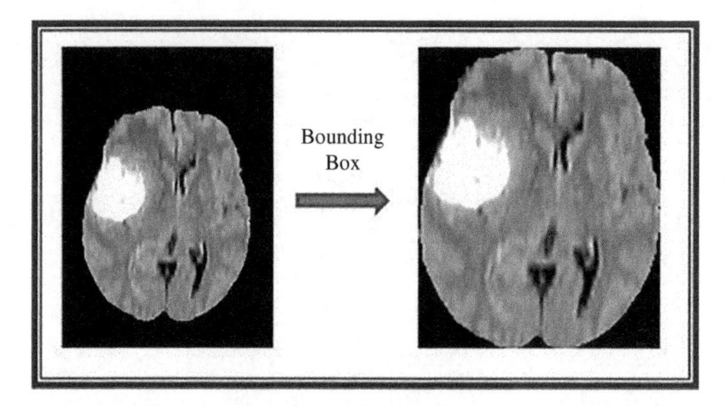

Figure 9.8 Original and resulting images with the bounding box.

3. Traverse column-wise through the input image from left to right until it reaches non-zero data. Assign the column value in *colmin*.
4. Traverse column-wise through the input image from right to left until it reaches non-zero data. Assign the column value in *colmax*.

This effectively crops the zero background and yields the AOI image within the *rowmin, rowmax, colmin* and *colmax* coordinates. A sample image of an MRI brain following AOI extraction is shown in Figure 9.8.

9.3.2 Feature Extraction

The input MRI image was split into two halves along the central bisecting line. As a result, the two left and right cerebral hemispheres (LHS and RHS, respectively) were obtained. Each hemisphere was treated as an individual image for feature extraction. The LHS and RHS images are shown in Figure 9.9. First-order, second-order, and third-order image features were computed using the image histogram, GLCM, and GLRLM, respectively. The Mean, Variance, Skewness, Kurtosis, and Energy were the features extracted from the histogram; they are tabulated in Table 9.1. Upto 20 features similar to Autocorrelation, Contrast, Correlation, and Cluster Prominence

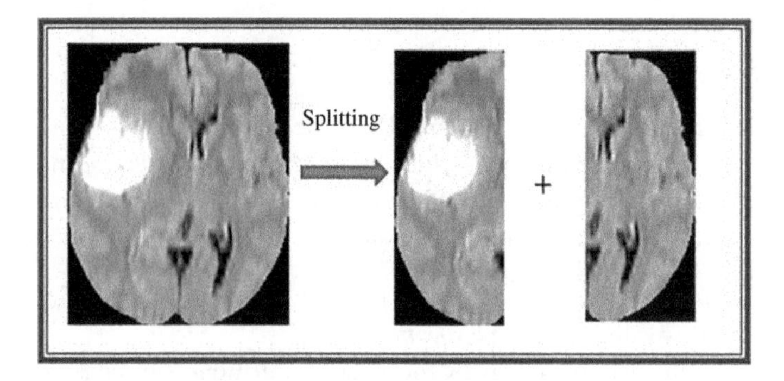

Figure 9.9 Splitting the whole image into LHS and RHS along the middle sagittal plane.

Table 9.1 Histogram-based image features.

SL. No	Feature Name	Formula
1	Mean	$P_1 = \sum_{x=lmin}^{lmax} x\,h(x)$
2	Variance	$P_2 = \sum_{x=lmin}^{lmax} (x - P_1)^2\, h(x)$
3	Skewness	$P_3 = \sum_{x=lmin}^{lmax} \frac{(x - P_1)^3\, h(x)}{(P_1)^{3/2}}$
4	Kurtosis	$P_4 = \sum_{x=lmin}^{lmax} \frac{(x - P_1)^4\, h(x)}{(P_1)^{4/2}}$
5	Entropy	$P_s = \sum_{x=lmin}^{lmax} h(x) \log_2 [h(x)]$

were extracted from the GLCM (see Table 9.2). Upto 11 features like Short Run Emphasis and Long Run Emphasis were extracted from the GLRLM (see Table 9.3). Additionally, the minimum, maximum, and mean intensity were extracted from the hemispheric images. The difference between the feature values were taken into consideration for the training and testing phase.

Normal images have smaller differences, whereas the tumor images will create larger differences between the LHS and RHS images. Finally, a feature set along with the class label was created with 763×39 dimensions for a collection of normal and abnormal images. The class values were labeled "1" for abnormal images and "0" for normal images.

The histogram-based features are listed in the Table 9.1. The value x denotes the intensity of a pixel; $lmin$ and $lmax$ are the minimum and maximum gray levels of the image, respectively. Function h(x) is the probability of x in the image.

The GLCM-based features are listed in Table 9.2. The **CM(i,j)** denotes the co-occurrence matrix, and N denotes the number of gray levels (eight in this study). $(CM_{x},)^{-}, (CM_{y},)^{-}, S_{x}$, and S_y are the mean and standard deviations of $\mathbf{CM}_x$ and $\mathbf{CM}_y$, respectively. The GLRLM features are listed along with their formulas in Table 9.3. The run length matrix is **RM(i,j)**. The number of runs is n_r; N^{-} is the number of gray levels; M^{-} is the size of pixels in the region of interest (Harshavardhan et al., 2017).

9.3.3 Feature Selection

Selecting the relevant features from the large feature set was important to improve the efficiency in terms of both computational cost and performance of the predictive model. The proposed method use the Pearson Correlation Coefficient, which is a filter-type feature selection method used for feature selection process.It is a

Table 9.2 GLCM-based image features.

Sl. No	Feature Name	Formula
1	Auto Correlation	$\sum_{i=1}^{N}\sum_{j=1}^{N}(i.j)\,\boldsymbol{CM}(i,j)$
2	Contrast	$\sum_{i=1}^{N}\sum_{j=1}^{N}\boldsymbol{CM}(i,j).(i-j)^2$
3	Correlation	$\sum_{i=1}^{N}\sum_{j=1}^{N}\frac{(i,j).\boldsymbol{CM}(i,j)-\overline{\boldsymbol{CM}_x\boldsymbol{CM}_y}}{S_xS_y}$
4	Cluster Prominence	$\sum_{i=1}^{N}\sum_{j=1}^{N}\left(i+j-2\overline{\boldsymbol{CM}}\right)^3\boldsymbol{CM}(i,j)$
5	Cluster Shade	$\sum_{i=1}^{N}\sum_{j=1}^{N}\left(i+j-2\overline{\boldsymbol{CM}}\right)^4\boldsymbol{CM}(i,j)$
6	Dissimilarity	$\sum_{i=1}^{N}\sum_{j=1}^{N}\lvert i-j\rvert\cdot\boldsymbol{CM}(i,j)$
7	Energy	$\sum_{i=1}^{N}\sum_{j=1}^{N}\boldsymbol{CM}(i,j)^2$
8	Entropy	$-\sum_{i=1}^{N}\sum_{j=1}^{N}\boldsymbol{CM}(i,j)\log\boldsymbol{CM}(i,j)$
9	Homogeneity	$\sum_{i=1}^{N}\sum_{j=1}^{N}\frac{\boldsymbol{CM}(i,j)}{1+(i-j)^2}$
10	Maximum Probability	$max_{i,j}\boldsymbol{CM}(i,j)$
11	Variance (sum of squares)	$\sum_{i=1}^{N}\sum_{j=1}^{N}\left(i-\overline{\boldsymbol{CM}}\right)^2\boldsymbol{CM}(i,j)$
12	Sum average (Sa)	$\sum_{i=2}^{2N}i\,\boldsymbol{CM}_{x+y}(i)$
13	Sum Variance (SV)	$\sum_{i=2}^{2N}\left(i-\boldsymbol{CM}_{x+y}\right)^2\boldsymbol{CM}_{x+y}(i)$
14	Sum Entropy (SE)	$-\sum_{i=2}^{2N}\boldsymbol{CM}_{x+y}(i)\,log\left\{\boldsymbol{CM}_{x+y}(i)\right\}$
15	Difference Variance	$\sum_{i=1}^{N}\left(i-\overline{\boldsymbol{CM}_{x-y}}\right)^2\boldsymbol{CM}_{x-y}(i)$
16	Difference Entropy	$-\sum_{i=1}^{N}\boldsymbol{CM}_{x-y}(i)\log\boldsymbol{CM}_{x-y}(i)$

Table 9.2 (*Continued*)

Sl. No	Feature Name	Formula
17	Information measure of Correlation	$\frac{HXY - HXY_1}{\max(HX, HY)}$ $HXY = -\sum_i \sum_j \boldsymbol{CM}(i,j) \log_2 \boldsymbol{CM}(i,j)$ HX-entropy of $\boldsymbol{CM}_x$, HY- entropy of $\boldsymbol{CM}_y$
18	Inverse Difference (INV)	$\sum_{i=1}^{N} \sum_{j=1}^{N} \frac{\boldsymbol{CM}(i,j)}{1 + \lvert i-j \rvert}$
19	Inverse Difference Normalized (INN)	$\sum_{i=1}^{N} \sum_{j=1}^{N} \frac{\boldsymbol{CM}(i,j)}{1 + \left(\frac{\lvert i-j \rvert}{N} \right)}$
20	Inverse Difference Moment Normalized	$\sum_{i=1}^{N} \sum_{j=1}^{N} \frac{\boldsymbol{CM}(i,j)}{1 + \left(\frac{\lvert i-j \rvert^2}{N^2} \right)}$

measure of the linear relationship between two attributes. It ranges from negative one to one. Negative one refers to a perfect negative correlation, and one refers to a perfect positive correlation. The formula for Pearson's correlation coefficient is given in Equation 9.4.

$$r = \frac{\sum_i (x_i - x^-)(y_i - y^-)}{\sqrt{\sum_i (x_i - x^-)^2} \sqrt{\sum_i (y_i - y^-)^2}}. \tag{9.4}$$

Here, x and y are the two attributes involved in the correlation calculation (Biesiada and Duch, 2007).

9.3.4 Classification

Classification is the process of identifying the class of a data after being trained by voluminous data. The performance of the classification model improves when it is trained using a large data set. There are two main types of classification techniques: binary (only two classes) and multi-class (more than two classes). The classifier used in this study was KNN.

9.3.5 Cross-Validation

Cross-validation is a way to estimate the accuracy of the predictive models by splitting the available data into training and testing data. It is also used to improve the accuracy or select the predictive model for the current scenario. Among various cross-validation techniques, the stratified k-fold cross-validation technique is used in the current work. Stratified k-fold cross-validation initially allows the training

Table 9.3 GLRLM-based image features.

Sl. No	Feature Name	Formula
1	Short Run Emphasis (SRE)	$\frac{1}{n_r}\sum_{i=1}^{N}\sum_{j=1}^{M}\frac{\boldsymbol{RM}(i,j)}{j^2}$
2	Long Run Emphasis (LRE)	$\frac{1}{n_r}\sum_{i=1}^{N}\sum_{j=1}^{M}\boldsymbol{RM}(i,j).j^2$
3	Gray level Non- Uniformity (GLNU)	$\frac{1}{n_r}\sum_{i=1}^{N}\left(\sum_{j=1}^{M}\boldsymbol{RM}(i,j)\right)^2$
4	Run length Non-Uniformity (RLNU)	$\frac{1}{n_r}\sum_{j=1}^{M}\left(\sum_{i=1}^{N}\boldsymbol{RM}(i,j)\right)^2$
5	Run Percentage (RP)	$\frac{n_r}{\sum_{i=1}^{N}\sum_{j=1}^{M}\boldsymbol{RM}(i,j).j}$
6	Low Gray Level Run Emphasis (LGRE)	$\frac{1}{n_r}\sum_{i=1}^{N}\sum_{j=1}^{M}\frac{\boldsymbol{RM}(i,j)}{i^2}$
7	High Gray Level Run Emphasis (HGRE)	$\frac{1}{n_r}\sum_{i=1}^{N}\sum_{j=1}^{M}\boldsymbol{RM}(i,j)\,i^2$
8	Short Run Low Gray Level Emphasis (SRLGE)	$\frac{1}{n_r}\sum_{i=1}^{N}\sum_{j=1}^{M}\frac{\boldsymbol{RM}(i,j)}{i^2.j^2}$
9	Short Run High Gray Level Emphasis (SRHGE)	$\frac{1}{n_r}\sum_{i=1}^{N}\sum_{j=1}^{M}\frac{\boldsymbol{RM}(i,j).i^2}{j^2}$
10	Long Run Low Gray Level Emphasis (LRLGE)	$\frac{1}{n_r}\sum_{i=1}^{N}\sum_{j=1}^{M}\frac{\boldsymbol{RM}(i,j).j^2}{i^2}$
11	Long Run High Gray Level Emphasis (LRHGE)	$\frac{1}{n_r}\sum_{i=1}^{N}\sum_{j=1}^{M}\boldsymbol{RM}(i,j).i^2.j^2$

data to be shuffled based on the class variable. It ensures that each fold has the same proportion of observations with a given class value. The training and validation are performed for k iterations by splitting the entire data into k-folds in each iteration. Each time, a different combination of training and testing data are formed such that one fold is allotted as validation data and the remaining are used as training data. Figure 9.10 shows a representation of stratified k-fold cross-validation for $k = 5$ (Kohavi, 1995; Rodriguez et al., 2010).

9.3.6 Training Validation and Testing

For training the model, six abnormal volumes from the BraTS 2013 data set and two normal volumes from the IBSR-18 data set were taken. In total, 763 images were used

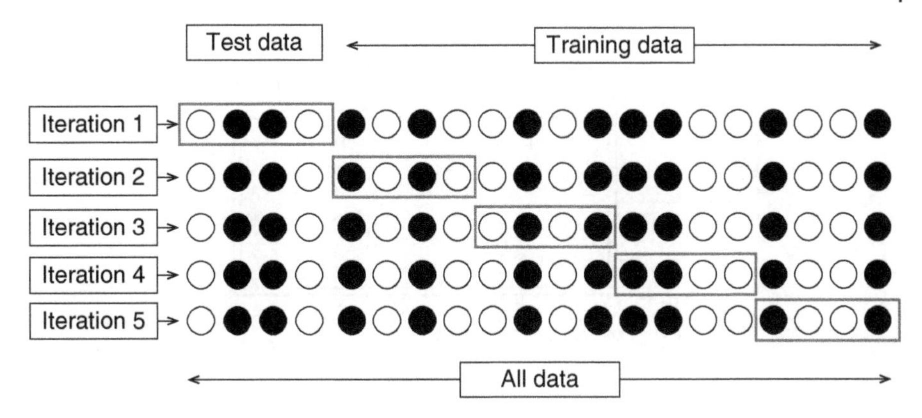

Figure 9.10 Stratified k-fold cross validation.

as training data. In validation, the training data was split into k folds for k iterations. In each iteration, one fold of data was used for validation and the remaining used as training data. To test the model, 11 tumor volumes from the BraTS 2013 and two normal volumes from the IBSR-18 data sets were used.

9.4 Materials and Metrics

The data used for the current experiment was taken from FLAIR sequences of the BraTS2013 data set (Menze et al., 2015; Kistler et al., 2013). The FLAIR images were chosen since the CSF is suppressed and tumor regions enhanced. A combination of normal and abnormal images were taken for training and testing. The model was implemented in a Python environment in a system with 2.6 GHz, 4 GB RAM and an Intel i5 processor. There were 763 images taken to train the model. The data set contained 256 normal and 507 abnormal images. The metrics used to examine the performance of the model were false alarm (FA), missed alarm (MS), and accuracy, which were computed from a confusion matrix.

9.4.1 Confusion Matrix

A confusion matrix summarizes the prediction results of the classification problems. The count of correct and incorrect predictions is placed in four classes as shown in Figure 9.11.

The formulas for false alarm, missed alarm, and accuracy are computed from the confusion matrix and as shown in Equations 9.5, 9.6, and 9.7 (Visa et al., 2011; Powers, 2011).

- ***False alarm:*** Normal slices are wrongly identified as tumor slices.

$$\%\ False\ Alarm\ (\mathrm{FA}) = \left(\frac{\mathrm{FN}}{Totalnumberofslices}\right)100 \tag{9.5}$$

Confusion Matrix		Actual Class	
		Positive (1)	Negative (0)
Prediction Class	Positive (1)	True Positive (TP)	False Positive (FP)
	Negative (0)	False Negative (FN)	True Negative (TN)

Figure 9.11 Structure of a confusion matrix.

- ***Missed alarm***: Tumor slices are wrongly identified as normal slices.

$$\%\ Missed\ Alarm\ (\mathrm{MA}) = \left(\frac{\mathrm{FP}}{Totalnumberofslices}\right)100 \tag{9.6}$$

- ***Accuracy***: The percentage of positive and negative results that were correctly detected.

$$Accuracy = \frac{\mathrm{TP}+\mathrm{TN}}{\mathrm{TP}+\mathrm{TN}+\mathrm{FP}+\mathrm{FN}} \tag{9.7}$$

9.5 Results and Discussion

The first-, second-, and third-order features were collected from the images and fed as input to the model. The feature set contains a collection of 39 attributes computed from the image histogram, GLCM, and GLRLM. Among the 39 features, 16 were selected by the Pearson's correlation coefficient with a correlation value greater than 0.3. This value was fixed by experimental study. The selected features and their correlation values are listed in Table 9.4. For training the KNN model, feature data with 763×16 dimensions were used. The KNN classifier was used as the learning algorithm in this model with a k value of 9. Stratified k-fold cross-validation was used to validate the classification model in $k = 5$ iterations. This phase resulted in a 98% accuracy of the model in an average of k iterations. The model was tested using 11 tumor volumes from the BraTS 2013 and two normal volumes from the IBSR-18 data sets. The slices in the middle of the tumor volumes were well predicted as abnormal, whereas the slices at the beginning and end of the volume were misclassified as abnormal because of the lack of symmetry between the LHS and RHS. The predictions made by the current model for the testing volumes are given in Table 9.5.

From Table 9.5, the normal volumes do not have any wrong predictions as all the slides show the LHS and RHS symmetry. Only selected slices from the tumor volumes were considered for testing. Slices with major tumor regions were predicted correctly. Slices at the beginning and end of the volumes may lack symmetry between the LHS and RHS and be predicted as abnormal, which categorizes them

Table 9.4 Correlation of the selected features with the class label.

Feature Name	Correlation with Class Label
Hist_Mean	0.3622
Histo_ Energy	0.368566
GLCM_Autocorrelation	0.344718
GLCM_Correlation	0.416017
GLCM_Cluster Prominence	0.335945
GLCM_Energy	0.373379
GLCM_Max_probability	0.459
GLCM_Sum of sqaures	0.334856
GLCM_Sum average	0.336314
GLCM_Sum entropy	0.393762
GLCM_Inforcorr	0.562027
GLCM_Informcorrelation	0.575247
GLRLM_SRE	0.357658
GLRLM_GLN	0.527891
GLRLM_RLN	0.610582
GLRLM_RunPercentage	0.442014
Class Label	1

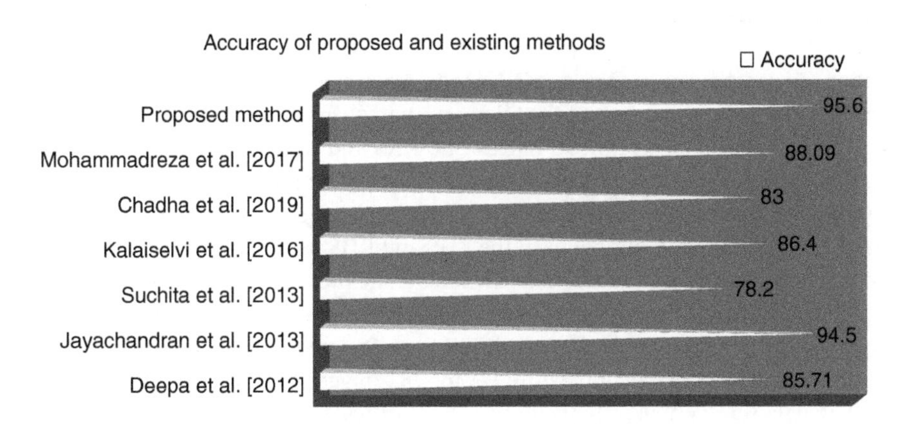

Figure 9.12 Accuracy of proposed and existing methods.

as false alarms (FA). Similarly, some abnormal slices with smaller tumors may fail to produce greater feature differences between the LHS and RHS and be predicted as normal, which categorizes them as missed alarms (MA). Based on the values of FA and MA, the accuracy of the current KNN model reached 96%. The comparison between the proposed and existing methods is shown in Figure 9.12 as the percent accuracy.

Table 9.5 Test phase predictions for the selected volumes from the IBSR-18 and BraTS 2013 data sets.

S. No	Volume Name	Total Slices	Abnormal Slices		FA (%)	MA (%)	Accuracy (%)
			Actual	Detected			
1	IBSR_10	125	–	–	0	0	100
2	IBSR_15	152	–	–	0	0	100
3	BHG7	126	54–113	57–111	0	4	96
4	BHG8	127	24–102	28–98	0	6.3	94
5	BHG9	118	34–106	39–105	0	4.2	96
6	BHG10	139	58–98	62–98	0	3.6	96
7	BHG11	93	28–95	32–96	1.1	4.3	94
8	BHG12	91	44–86	46–82	3.3	6.6	90
9	BHG13	78	49–79	50–77	0	6.4	94
10	BHG14	85	10–82	11–80	0	3.5	97
11	BHG15	95	16–95	18–94	0	3.2	97
12	BHG22	69	6–62	4–59	3	4.4	93
13	BLG6	48	6–44	5–45	4.1	0	96
Average		–			**.96**	**3.6**	**95.6**

9.6 Conclusion

In the present study, we proposed a KNN model to predict whether MRI brain images are normal or abnormal. It used the bilateral symmetric nature of the human brain to split the input image into two halves, the LHS and RHS, based on the middle sagittal line. The first-order, second-order and third-order image features of the histogram, GLCM, and GLRLM, respectively, were extracted from both the LHS and RHS images. The difference between the LHS and RHS features were used to make the prediction, since greater differences indicate the presence of an abnormality in either of the hemispheres. There were 39 features extracted, from which 16 features were selected by the Pearson's correlation coefficient to have correlation values greater than 0.3. The selected feature set for 763 images from the BraTS2013 and IBSR-18 data sets were used for training. The features of two normal volumes from the IBSR-18 and 11 abnormal volumes from the BraTS2013 data sets were used for testing. The model reached an accuracy of 95.6%, which is competitive with state-of-the-art methods.

References

F. Albregtsen. Statistical texture measures computed from gray level coocurrence matrices. 2008. Corpus ID: 18927855.

Jacek Biesiada and Wlodzisław Duch. Feature selection for high-dimensional data — a pearson redundancy based filter. pages 242–249. Springer Berlin Heidelberg, 2007. doi: 10.1007/978-3-540-75175-5_30.

K. ChandraPrabha. Texture analysis using GLCM & GLRLM feature extraction methods. *International Journal for Research in Applied Science and Engineering Technology*, 7(5):2059–2064, May 2019. doi: 10.22214/ijraset.2019.5344.

S.N. Deepa and B.A. Devi. Artificial neural networks design for classification of brain tumour. In *2012 International Conference on Computer Communication and Informatics*. IEEE, January 2012. doi: 10.1109/iccci.2012.6158908.

S. Goswami and L.K.P. Bhaiya. A hybrid neuro-fuzzy approach for brain abnormality detection using GLCM based feature extraction. In *International Conference on Emerging Trends in Communication, Control, Signal Processing and Computing Applications (C2SPCA)*, pages 1–7. IEEE, October 2013. doi: 10.1109/c2spca.2013.6749454.

A. Harshavardhan, S.S. Babu, and T. Venugopal. Analysis of feature extraction methods for the classification of brain tumor detection. *International Journal of Pure and Applied Mathematics*, 117(7):145–155, 2017. Online ISSN: 1314-3395.

A. Jayachandran and R. Dhanasekaran. Brain tumor detection and classification of mr images using texture features and fuzzy SVM classifier. *Research Journal of Applied Sciences, Engineering and Technology*, 6(12): 2264–2269, July 2013. doi: 10.19026/rjaset.6.3857.

T. Kalaiselvi and S.K. Selvi. Energy update restricted chan–vese model for tumor extraction from MRI of human head scans. *International Journal of Computational Methods*, 15(01):1750081, September 2017. doi: 10.1142/s0219876217500815.

S. Karuppanagounder and K. Thiruvenkadam. A novel technique for finding the boundary between the cerebral hemispheres from MR axial head scans. In B. Prasad, P. Lingras, and A. Ram, editors, *Proceedings of the 4th Indian International Conference on Artificial Intelligence, IICAI 2009, Tumkur, Karnataka, India, December 16-18, 2009*, pages 1486–1502. IICAI, 2009.

J. Kim, B.S. Kim, and S. Savarese. Comparing image classification methods: K-nearest-neighbor and support-vector-machines. In *Proceedings of the 6th WSEAS International Conference on Computer Engineering and Applications, and Proceedings of the 2012 American Conference on Applied Mathematics*, AMERICAN-MATH'12/CEA'12, pages 133–138, Stevens Point, Wisconsin, USA, 2012. World Scientific and Engineering Academy and Society (WSEAS). ISBN 9781618040640. doi: 10.5555/2209654.2209684.

M. Kistler, S. Bonaretti, M. Pfahrer, R. Niklaus, and P. Büchler. The virtual skeleton database: An open access repository for biomedical research and collaboration. *Journal of Medical Internet Research*, 15(11):e245, November 2013. doi: 10.2196/jmir.2930.

R. Kohavi. A study of cross-validation and bootstrap for accuracy estimation and model selection. In *Proceedings of the 14th International Joint Conference on Artificial Intelligence - Volume 2*, IJCAI'95, pages 1137–1143, San Francisco, CA, USA, 1995. Morgan Kaufmann Publishers Inc. ISBN 1558603638. doi: 10.5555/1643031.1643047.

H. Le, T. Le, S. Tran, H. Tran, and N. Thuy. Image classification using support vector machine and artificial neural network. *International Journal of Information Technology and Computer Science*, 4:32–38, 05 2012. doi: 10.5815/ijitcs.2012.05.05.

C. Megha and J. Sushma. Detection of brain tumor using machine learning approach. pages 188–196. Springer Singapore, 2019. doi: 10.1007/978-981-13-9939-8_17.

B.H. Menze, A. Jakab, S. Bauer, J. Kalpathy-Cramer, K. Farahani, and et al. The multimodal brain tumor image segmentation benchmark (BRATS). *IEEE Transactions in Medical Imaging*, 34(10):1993–2024, October 2015. doi: 10.1109/tmi.2014.2377694.

D. Nader, H. Jehlol, A. Subhi, and A. Oleiwi. Brain tumor detection using shape features and machine learning algorithms. *International Journal of Scientific & Engineering Research*, 6(12):454–459, December 2015.

D. Powers. Evaluation: From precision, recall and f-factor to roc, informedness, markedness & correlation. *Journal of Machine Learning Technologies*, 2(1): 37–63, 01 2011. ISSN: 2229-3981.

R. Ramteke and M. Khachane. Automatic medical image classification and abnormality detection using k- nearest neighbour. *International Journal of Advanced Computer Research*, 2(4):190–196, 12 2012.

A. Renjith, P. Manjula, and P.M. Kumar. Brain tumour classification and abnormality detection using neuro-fuzzy technique and Otsu thresholding. *Journal of Medical Engineering & Technology*, 39(8):498–507, October 2015. doi: 10.3109/03091902.2015.1094148.

J.D. Rodriguez, A. Perez, and J.A. Lozano. Sensitivity analysis of k-fold cross validation in prediction error estimation. *IEEE Transactions on Pattern Analysis and Machine Intelligence*, 32(3):569–575, March 2010. doi: 10.1109/tpami.2009.187.

M. Soltaninejad, G. Yang, T. Lambrou, N. Allinson, T.L. Jones, and et al. Automated brain tumour detection and segmentation using superpixel-based extremely randomized trees in FLAIR MRI. *International Journal of Computer Assisted Radiology and Surgery*, 12(2):183–203, September 2016. doi: 10.1007/s11548-016-1483-3.

M.B. Stegmann, K. Skoglund, and C. Ryberg. Mid-sagittal plane and mid-sagittal surface optimization in brain MRI using a local symmetry measure. volume 5747, pages 568–579. SPIE, April 2005. doi: 10.1117/12.595222.

Kalaiselvi T., Kumarashankar P., Sriramakrishnan P., and Nagaraja P. Novel statistical feature for brain abnormality detection process in image mines of MRI head volumes. *Recent Advances in Computer Science and Applications*, pages 119–123, December 2015. ISBN: 978-93-84743-57-4.

K. Thiruvenkadam. *Brain Portion Extraction and Brain Abnormality Detection from Magnetic Resonance Imaging of Human Head Scans*. Pallavi Publications South India Pvt Ltd., 2011.

K. Thiruvenkadam, K. Nagarajan, and S.n Padmanaban. Automatic brain tissues segmentation based on self initializing k-means clustering technique. *International Journal of Intelligent Systems and Applications*, 9(11):52–61, November 2017. doi: 10.5815/ijisa.2017.11.07.

Kalaiselvi Thiruvenkadam, Sriramakrishnan Padmanaban, and Somasundaram Karuppanagounder. Brain abnormality detection from MRI of human head scans using the bilateral symmetry property and histogram similarity measures. In *The 20th International Conference on Computer Science and Engineering 2016 IEEE Explorer.* IEEE, December 2016. doi: 10.1109/ICSEC.2016.7859867.

S. Visa, B. Ramsay, A.L. Ralescu, and E. van der Knaap. Confusion matrix-based feature selection. In *MAICS*, volume 710, pages 120–127, January 2011.

C.C. Yang, S.O. Prasher, P. Enright, C. Madramootoo, M. Burgess, and et al. Application of decision tree technology for image classification using remote sensing data. *Agricultural Systems*, 76(3):1101–1117, 2003. ISSN 0308-521X. doi: https://doi.org/10.1016/S0308-521X(02)00051-3. URL https://www.sciencedirect.com/science/article/pii/S0308521X02000513.

X. Zhou, S. Wang, W. Xu, G. Ji, P. Phillips, P. Sun, and Y. Zhang. *Detection of pathological brain in MRI scanning based on wavelet-entropy and naive Bayes classifier*. volume 9043, pages 201–209. Springer International Publishing, 2015. doi: 10.1007/978-3-319-16483-0_20.

10

Conclusion

*Siddhartha Bhattacharyya**

CHRIST (Deemed to be University), Bengaluru, India

With the sudden passage of deadly diseases across the world, medical practitioners have come under tremendous pressure to curb the casualties due to the inherent limitations of healthcare resources at their disposal. Even though high-income nations are better equipped with the latest medical infrastructure, they often feel helpless with the suddenness of disease influx. Thus, the precarious situation in middle- and low-income nations is a matter of grave concern.

Given the limited healthcare resources and infrastructure, added to the ever-increasing population of low-income nations, a well-framed policy on the optimized use of the scarce healthcare resources is the need of the hour. The necessity of an optimized allocation of healthcare resources has been well substantiated with the emergence of several deadly and infectious diseases. In addition, the requirement for non-contact testing facilities and effective patient care systems have also assumed paramount importance.

The field of healthcare informatics (Sami, 2021; Shortliffe and Cimino, 2014; Patel and Kannampallil, 2015) have evolved of late, thanks to computational scientists across the globe. Healthcare informatics refers to an effective way of managing healthcare information with the objective of efficiently catering patient healthcare information toward equitable healthcare opportunities. The primary aim of such an informatics system is to optimize the ethical acquisition, storage, retrieval, and use of information in health and bio-medicine to derive a justified fail-safe solution to the supply-demand paradigm. It is quite clear that the management of such huge amounts of patient-specific data and information, often to the tune of terabytes, is often an unfathomable task. The extraction and use of relevant diagnostic data pose serious challenges to computational scientist. Moreover, redundancy in the data is also a matter of concern that not only introduces complexity, but also uncertainty and imprecision.

Intelligent healthcare informatics (also referred to as smart healthcare informatics) augments the purview of the existing health care amenities by enveloping

*dr.siddhartha.bhattacharyya@gmail.com

Applied Smart Health Care Informatics: A Computational Intelligence Perspective, First Edition.
Edited by Sourav De, Rik Das, Siddhartha Bhattacharyya, and Ujjwal Maulik.

intelligent technologies toward information engineering aspects. Intelligent analysis of the information therein enhances the overall management, as far as the use of resources is concerned. By means of a multidisciplinary approach, it uses health information technology, while resorting to computational intelligence, to improve health care by resorting to newer and higher quality opportunities.

Thus, the advent of intelligent healthcare informatics (Quan and Sanderson, 2018; Huser and Shmueli-Blumberg, 2018) has unfurled a new era of equitable healthcare opportunities while doing away with waste and under-use of healthcare resources. The touch of computational intelligence has revolutionized this field with the evolution of intelligent early prediction, diagnostic, and prognostic procedures, thereby reducing the effective downtime. Use of computational intelligence has also ushered in time-efficient, accurate, and fail-safe techniques compared to analog, manual processes. Even testing facilities have become more agile and non-destructive in nature, thanks to the use of intelligent tools and methods.

Another far reaching effect of smart healthcare informatics is the evolution of telemedicine facilities, where the main aim is accurate diagnosis, instead of hovering, and investing time and effort to explore means of distributing information. Development in communication has eased the data transmission process to such an extent that considerable time can be saved before critical healthcare support is needed. Intelligent healthcare informatics have also facilitated the processing and analysis of huge patient-specific data in almost no time, thereby reducing the threat of higher casualty rates.

With the advent of Big Data analysis, intelligent health care informatics has been calling for efficient and effective use of healthcare data and the diagnosis thereof. During the next few years, there must be a sea change in the approaches to health care management. Smart pills may come to the foray as Bio-MEMS or intelligent drug delivery systems. Wearable medical devices could be attached to the patient's body to allow real-time monitoring by physicians (Huser and Shmueli-Blumberg, 2018). Nano-bots may be introduced into the body to collect specimens or look for early signs of diseases. Content management might also become more intelligent and intricate.

This volume is, thus, a novel initiative to report the latest innovations in healthcare informatics with reference to computational intelligence-assisted modalities and techniques. The contributory chapters focus on the innovative perspectives in smart healthcare management and information analysis. Thus, this volume will surely come in good stead as a handy treatise for medical practitioners and related professionals.

References

V. Huser and D. Shmueli-Blumberg. Data sharing platforms for de-identified data from human clinical trials. *Clinical Trials*, 15(4):413–423, April 2018. doi: 10.1177/1740774518769655.

V.L. Patel and T.G. Kannampallil. Cognitive informatics in biomedicine and healthcare. *Journal of Biomedical Informatics*, 53:3–14, February 2015. doi: 10.1016/j.jbi.2014.12.007.

X.I. Quan and J. Sanderson. Understanding the artificial intelligence business ecosystem. *IEEE Engineering Management Review*, 46(4):22–25, December 2018. doi: 10.1109/emr.2018.2882430.

H. R. Sami. Medical informatics in neurology: What is medical informatics?, 2021.

E.H. Shortliffe and J.J. Cimino, editors. *Biomedical Informatics*. Springer London, 2014. doi: 10.1007/978-1-4471-4474-8.

Index

Applied Smart Health Care Informatics: A Computational Intelligence Perspective, First Edition.
Edited by Sourav De, Rik Das, Siddhartha Bhattacharyya, and Ujjwal Maulik.